Key Concepts STEM

SCIENCE TECHNOLOGY ENGINEERING MATH

PHYSICS

MECHANICS

General Editor: Anne O'Daly

BROWN BEAR BOOKS

Published by Brown Bear Books Ltd
4877 N. Circulo Bujia, Tucson, AZ 85718
USA
and
Studio G14, Regent Studios, 1 Thane Villas,
London N7 7PH, UK

Cataloging-in-Publication data available upon request.

ISBN 978-1-83572-082-0 (ALB)
ISBN 978-1-83572-087-5 (paperback)
ISBN 978-1-83572-092-9 (ebook)

Children's Publisher: Anne O'Daly
Design Manager: Keith Davis
Picture Manager: Sophie Mortimer
Consultant: Chris Woodford

Manufactured in the United States of America
CPSIA compliance information: Batch#AG/5665

Picture Credits
Front Cover: Shutterstock: Volodymyr Krasyuk: **Interior: Alamy:** Photos 12 34; NASA: 12–13, KSC 21; **Public Domain:** A Hodges 43, nagyman 45; **Shutterstock:** Mark Agnor 40–41, all about people 8, Belish 38, Burban 26, Colin Dewar 1, 32, Artur Didyk 51, Maruicio Graiki 24–25, Evren Kalinbacak 6, Sebastian Kaulitzki 4, LianeM 47r, Oleksandr Lutsenko 10, Master1305 47l, matimix 29t, Eugene Onischenko 28, RD Production 22, RobSt 48, Andreas Rose 56-57, Saint Fuangnakhon 52-53, Richard Semik 58-59, Juri Semjonow 55, V_E 39, Valley Journal 37, Zbynak 50; **U.S. Government:** NIST 7, United States Department of Energy 18.

All other artwork and photography © Brown Bear Books Ltd

Brown Bear Books has made every attempt to contact the copyright holder. If you have any information about omissions please contact licensing@brownbearbooks.co.uk.

STEM

Key Concepts in STEM: Physics describes the key theories in the study of physics and their practical applications. Much of physics concerns the movement of objects, and in order for something to move a force must be involved. The branch of physics that deals with forces and movement is called mechanics, and this is the primary subject of this book. Explanatory diagrams and informative photographs, detailed features on related aspects of the topics covered and the main scientists involved in the advancement of physics, and definitions of key "Science Words" all enhance the coverage. "Try This" sidebars outline experiments that learners can follow as a first step to practical investigations.

Contents

MEASURING MATTER

Measurement is at the heart of physics. Observation and measurement are central to the whole of science. Measurement requires units to express how heavy, long, or old something is. Science uses a wide range of units to measure everything from the size of an atom to the age of the Universe.

In everyday life a variety of units is used, usually chosen to suit the thing that is being measured. For example, the distance to the next town or city is measured in miles, the size of a parking lot in yards, the height of a flagpole in feet, and the size of a piece of paper in inches. These are all units of length: distance is simply a length along the ground, and height is a length measured in an upward direction.

But such a mixture of units can be a nuisance, particularly in the math involved in changing inches to feet or yards to miles. Also, the miles used in the United States may be different from the miles used in Finland or in China. North American pints and gallons are different from British pints and gallons. For example, in the U.S., 1 pint equals 0.473 liters, while in Britain, 1 pint equals 0.568 liters

These sausage-shaped objects are Salmonella *bacteria. Each is about a micrometer long. Here they are magnified 100,000 times in an artificially colored image from a scanning electron microscope (SEM).*

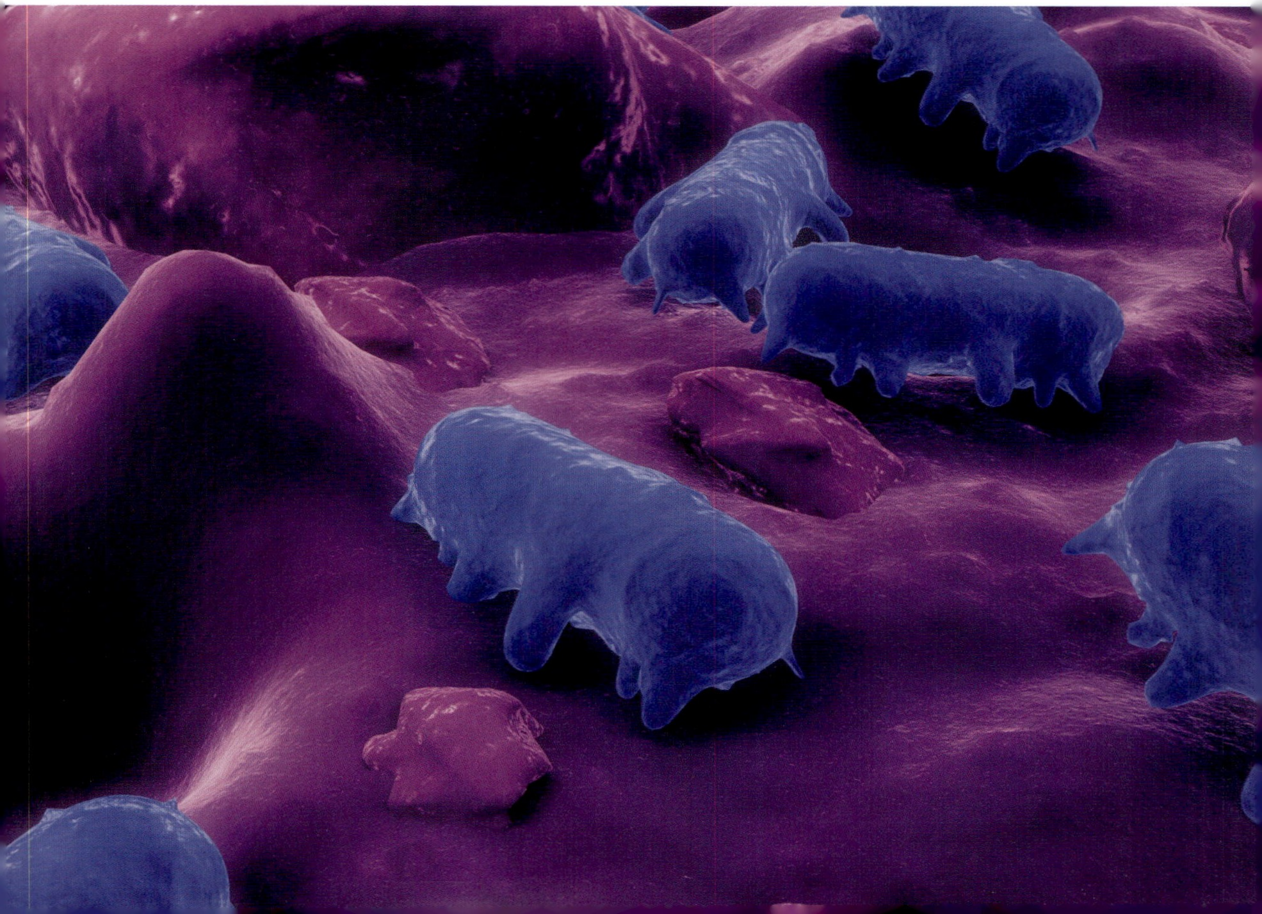

INTERNATIONAL SYSTEM OF UNITS

Shown below are the seven base units of the international system of units called SI (*Système International d'Unités*). Among the important derived units (combined base units) are the hertz, used for measuring frequency; the newton, for measuring force; the ohm, volt, and watt used for measuring electrical resistance, voltage, and power, respectively; and the joule, for measuring energy.

	Mass	Amount of substance	Length	Current	Luminous intensity	Temperature	Time
SI unit	kilogram	mole	meter	ampere	candela	kelvin	second
Symbol	kg	mol	m	A	cd	K	s

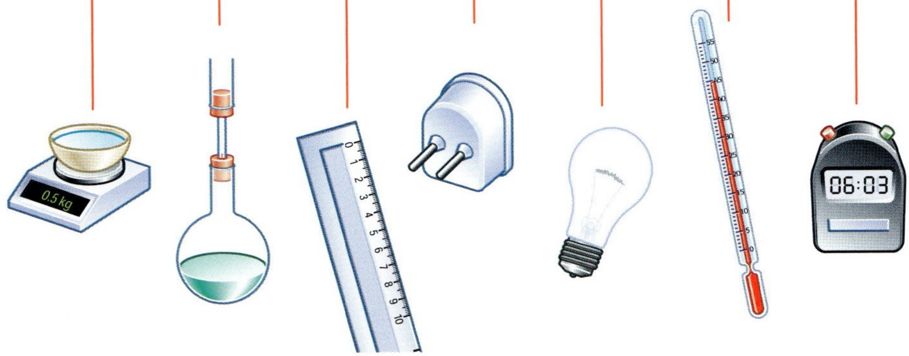

(the British pint is bigger than the U.S. pint). The British gallon is also 1.2 times larger than the U.S. gallon. Scientists get around these problems by having only one unit for length—the meter. Every length is measured in meters or in multiples of meters (such as kilometers) or in submultiples of meters (for example, centimeters).

A number of standard prefixes go before the word for the base unit to make the multiples and submultiples. For example, the prefix *kilo-* means "1,000 times:" 1 kilometer = 1,000 meters (written as 1 km = 1,000 m). In a similar way, *centi-* means "1/100:" 1 centimeter (1 cm) = 1/100 meter (0.01 m). Thus the distance from Chicago to Los Angeles is about 1,740 km; the length of a new pencil is about 18 cm.

SCIENCE WORDS

Luminous intensity:
The light-emitting power of a source of light.
Mass:
The amount of matter in an object.
SI units:
System of units used internationally in science (short for *Système International d'Unités*, its name in French). There are seven base units (ampere, candela, kelvin, kilogram, meter, mole, and second) and various derived units, which are combinations of base units.
Standard form:
A way of expressing very large or very small numbers that uses an index to represent powers of 10. For example: 10,000,000 is 10^7 and 0.000002 is $2 * 10^{-6}$.

Metric System

The meter is a unit in the metric system. This system was invented in France about 200 years ago, when the meter was taken to be a ten-millionth of the distance from the Equator to the North Pole along the Paris meridian. The kilogram is also a metric unit. The metric system is used for everyday measurements in most European countries and is becoming increasingly common in the United States.

SI System

Science uses a version of the metric system called the SI system (so called after its French name, *Système International d'Unités*). This

A pole vaulter clears the bar at a track-and-field contest. The achievements of athletes are measured in metric units. The world women's pole vault record is more than 5 meters (16.4 feet).

ANCIENT MEASURING INSTRUMENTS

This illustration shows some ancient types of measuring instrument: **(a)** human forearm, about 0.5 meters long; **(b)** simple balance for weighing; **(c)** water clock for telling the time; **(d)** sundial for telling the time of day; **(e)** astrolabe for measuring the angles of the stars; **(f)** hourglass for measuring elapsed time; **(g)** micrometer for measuring small thicknesses; **(h)** sextant for measuring the Sun's angle in the sky.

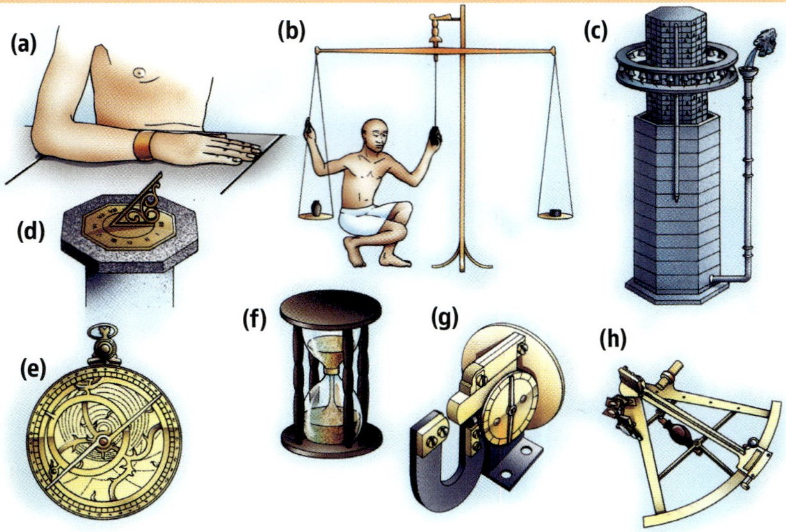

METRIC PREFIXES

Prefix	Symbol	Multiple	Prefix	Symbol	Multiple	Prefix	Symbol	Multiple
atto-	a	* 10^{-18}	centi-	c	* 10^{-2}	mega-	M	* 10^{6}
femto-	f	* 10^{-15}	deci-	d	* 10^{-1}	giga-	G	* 10^{9}
pico-	p	* 10^{-12}	deca-	da	* 10	tera-	T	* 10^{12}
nano-	n	* 10^{-9}	hecto-	h	* 10^{2}	peta-	P	* 10^{15}
micro-	µ	* 10^{-6}	kilo-	k	* 10^{3}	exa-	E	* 10^{18}
milli-	m	* 10^{-3}						

Here are some examples:
picofarad (pF), equal to 10–12 farads, used to measure capacitance
nanometer (nm), equal to 10–9 meters, used to measure molecules
microampere (µA), equal to 10–6 amperes, used to measure nerve impulses
milligram (mg), equal to 10–3 grams, used to weigh out medicines
centiliter (cl), equal to 10–2 liters, used to measure wine
hectare (ha), equal to 102 ares, used for areas of fields
kilovolt (kV), equal to 103 volts, used for railroad voltages
megawatt (MW), equal to 106 watts, used for a power-plant output
gigabyte (Gb), equal to 109 bytes, used for computer storage capacity

The U.S. National Institute of Standards and Technology (NIST) maintains the atomic clocks used in the United States. An atomic clock keeps time to an accuracy of better than 1 second in 30,000 years.

system has seven base units, shown at the top of page 5, two supplementary, and various derived units. There are 18 derived units, each with a special name and made from combinations of the seven base units. The base unit of mass is the kilogram (= 1,000 grams), chosen because the gram (about one thirthieth of an ounce) is too small for many measurements. In this book measurements are usually given in both SI and their customary equivalents, but sometimes only in SI units.

Standard Form

When measurements are made using SI or metric units, some of the numbers become very large indeed. For example, the Earth lies about 150 million km from the Sun, which in figures is 150,000,000 km. Standard form uses an index to express large numbers as powers of 10. For instance, 1,000 = 10^3 and 1,000,000 = 10^6. So the distance to the Sun is 1.5 * 10^8 km. A human hair is about a ten-thousandth of a meter across, or 0.0001 m. In standard form this is written as 1 * 10^{-4} m.

MEASURING TEMPERATURE

Temperatures are measured using a thermometer, a device that utilizes some property of a substance that changes when the substance is heated. The liquid-in-glass thermometer makes use of the fact that a liquid expands when it is heated. Another type makes use of the expansion of a metal when it is heated.

To express an object's level of hotness, a temperature scale and a temperature-measuring device are needed. The familiar mercury thermometer is a long, narrow glass tube with a bulb at one end that contains mercury. The other end of the tube is sealed. When the bulb is heated (by being placed where the temperature is to be measured), the mercury inside the tube expands and

SCIENCE WORDS

Bimetallic strip: A strip consisting of two different metals bonded together. The metals expand at different rates when they are heated, so the strip bends when it is heated.
Celsius scale: A temperature scale that has 100 degrees between the freezing point of water (0°C) and the boiling point of water (100°C). It used to be called the centigrade scale.
Fahrenheit scale: A temperature scale that has 180 degrees between the freezing point of water (32°F) and the boiling point of water (212°F).

Because mercury is a highly toxic element, many countries have banned mercury-in-glass thermometers for medical use. Instead, patients' temperatures are now taken using digital thermometers with liquid crystal displays (LCD), like the one shown below.

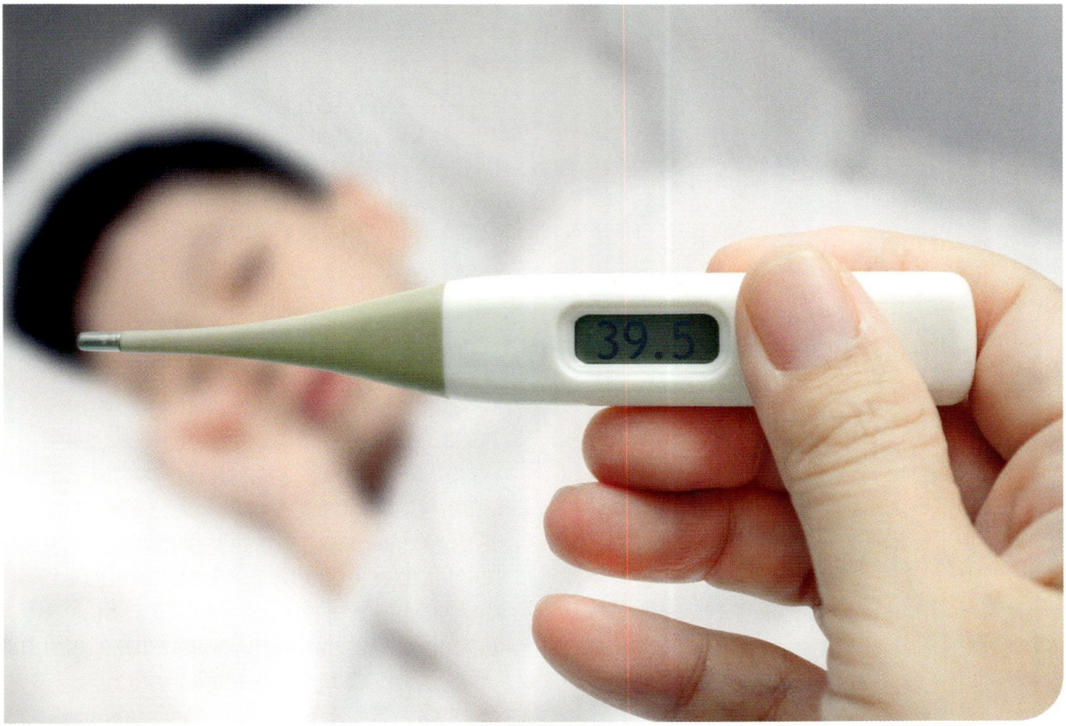

TYPES OF THERMOMETER

A liquid-in-glass thermometer relates temperature to the movement of a column of liquid expanding in a narrow glass tube. A thermocouple thermometer is based on two junctions between different metals held at different temperatures producing an electric voltage. Temperature in a platinum resistance thermometer is measured in terms of the electrical resistance of a platinum wire coil. A digital thermometer gives a direct temperature reading.

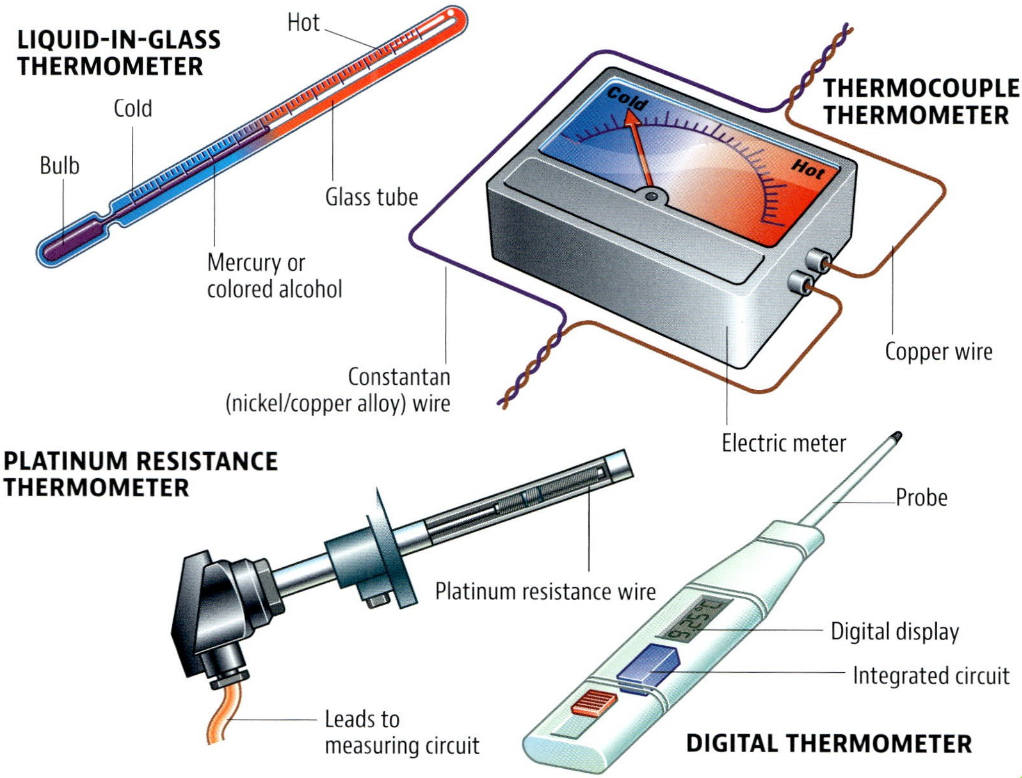

moves along the tube. The distance it moves gives a measure of the temperature. The glass tube is graduated in degrees, either in Fahrenheit or Celsius.

A clinical thermometer is a type of thermometer used for taking a person's temperature. It works over only a narrow range, usually 95 to 108°F (35 to 42°C). A person's "normal" body temperature is 98.6°F (37°C). There is a narrow kink in the tube just above the bulb, so when the thermometer is removed from the person's mouth, the mercury thread in the tube breaks at the kink. This leaves the upper end of the thread in position, so that the thermometer can be read. The mercury thread has to be shaken back into the bulb before the thermometer can be used again.

Because mercury freezes solid at a temperature of −38°F (−38.9°C), a mercury thermometer cannot be used below this temperature. Alcohol does not freeze until the temperate drops to −179°F (−117.3°C), and so it is suitable for low-temperature thermometers.

Expanding Metals

Liquid-in-glass thermometers make use of the expansion of liquids when they are heated. Metals also expand slightly when heated, but it is hard to make a mechanism that responds to the slight movement that results. The answer is to use a bimetallic strip—two different metals bonded together. Steel and brass are often used because their expansions are very different. When such a strip is heated, the brass expands more than the steel. This causes the strip to bend, and such movement is made use of in some thermostats, which are devices that maintain a constant temperature by controlling the flow of fuel or electricity to a heater.

In this type of thermometer, the bimetallic strip is a spiral that is fixed at one end. When the strip is heated, the spiral unwinds slightly. The movement is enough to move a pointer across a scale calibrated in degrees.

TWO-METAL SPIRAL

In this thermometer, strips of steel and brass are bonded together to form a bimetallic strip. The strip is wound into a coil and fixed at one end. As the temperature rises, the brass expands more than the steel, and the strip uncoils slightly, making a pointer move around a temperature scale.

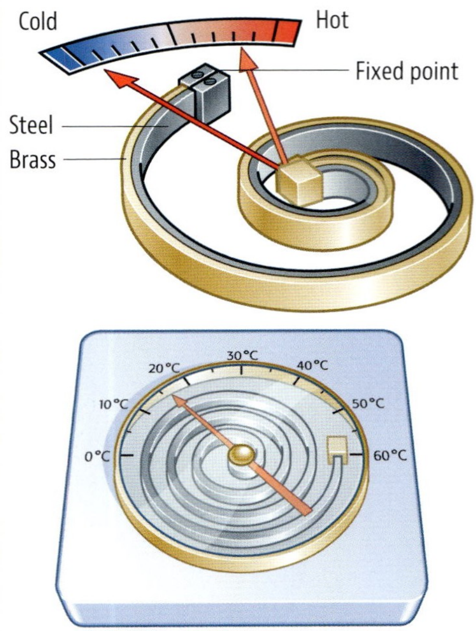

Other Types of Thermometer

In theory, any physical property that alters with a change in temperature can be used to make a thermometer. For example, in a thermocouple thermometer a junction between a pair of wires made from different metals is kept at a different temperature from another similar junction. This produces a small electric voltage in the wires. The voltage is measured with a sensitive voltmeter calibrated (marked out) directly in degrees to give a measure of temperature. Another electrical type is the platinum resistance thermometer. It contains a circuit that measures the change in electrical resistance of a platinum wire as its temperature changes. The resistance can then be matched to the actual temperature.

A thermostatic radiator valve controls the flow of hot water to radiators. It has a valve that opens and closes to control the hot water flow and a sensor that controls the opening of the valve.

Temperature Conversions

It is sometimes necessary to convert Celsius temperatures to Fahrenheit ones, or the other way around—Fahrenheit to Celsius. This

HIGH TEMPERATURES

Very high temperatures, such as those in a furnace, are measured using a pyrometer. In this device, the electric current flowing through a wire filament is adjusted until it has the same brightness as the furnace when viewed through the eyepiece. The meter measures the current, but is marked in degrees.

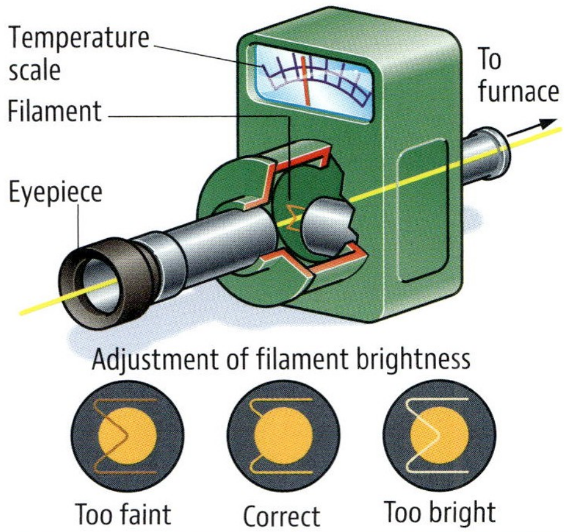

involves a little simple math. To change Celsius to Fahrenheit, multiply by 9 and divide by 5 (or just multiply once by 1.8), then add 32. To convert 50°C to Fahrenheit:

$$\frac{50 * 9}{5} = 90, \quad 90 + 32 = 122°F$$

To convert Fahrenheit to Celsius, reverse the procedure. First subtract 32, then multiply by 5 and divide by 9 (or just multiply once by 0.56). To convert 86°F to Celsius:

$$86 - 32 = 54, \quad \frac{54 * 5}{9} = 30°C$$

To convert Celsius to Kelvin (a scale used in really hot or cold places, such as in space) add 273. For Kelvin to Celsius, subtract 273. To convert Fahrenheit to Kelvin first convert from Fahrenheit to Celsius, and then add 273. Reverse the process for Kelvin to Fahrenheit.

SCIENCE WORDS

Thermocouple: Temperature-measuring device comprising two wires of different metals joined at their ends. When the joins are at different temperatures, an electric current flows in the wires and can be measured with a sensitive voltmeter, which displays the temperature difference between the joins.
Thermostat: A device for keeping a constant temperature, consisting of a temperature-sensitive element that controls the electricity or fuel supply to a heater.

MAXIMUM AND MINIMUM

This type of thermometer, used for weather records, shows the highest and lowest temperatures over a period of time. As the temperature rises, the alcohol in the right-hand bulb expands and the mercury rises in the left-hand tube. The mercury pushes a sprung steel marker, which stays in place when the temperature falls and the alcohol contracts. The mercury then rises in the right-hand tube, pushing a second marker. In this way, the markers show the maximum and minimum temperatures.

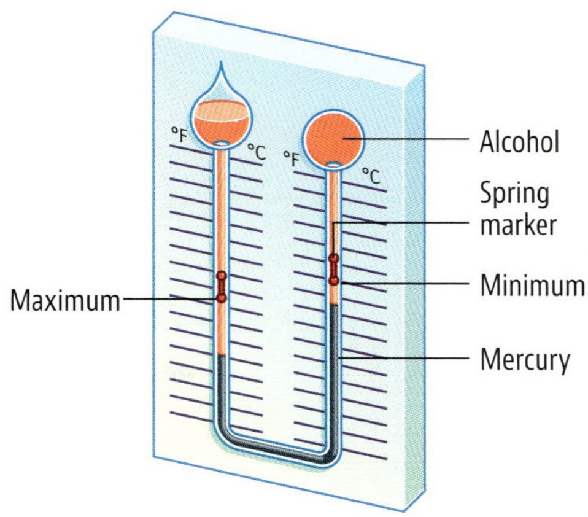

ATOMIC STRUCTURE, ATOMS, AND ISOTOPES

All matter is made from atoms—particles so small that about 100 million atoms can span a fingernail. There are some 5,000 trillion trillion atoms in the human body. In 1897, the English physicist J. J. Thomson discovered that atoms contain even smaller particles: called electrons.

Electrons have negative electric charge. There also has to be positive charge in the atom to balance these negative charges. Thomson suggested that the electrons were embedded in a globe of positive electricity, like raisins in a plum pudding. But then Ernest Rutherford (1871–1937), a New Zealander, found that the atom's positive charge and most of its mass lies at its center in a core called the nucleus. The atom is held together by the electrical attraction of the positive nucleus for the negative electrons.

Later, scientists found that the nucleus is itself made up of particles, or nucleons. They are of two types. One is the proton, which has

EARLY MODELS OF THE ATOM

Early theorists, such as John Dalton (1766–1844), believed the atom had no structure. In 1897, J. J. Thomson (1856–1940) found it actually contained negatively charged electrons. Ernest Rutherford (1871–1937) showed that the balancing positive charge was concentrated in a nucleus, and Niels Bohr (1885–1962) believed that electrons followed fixed paths or orbits. Scientists now believe that electrons occupy fuzzy "clouds" called orbitals.

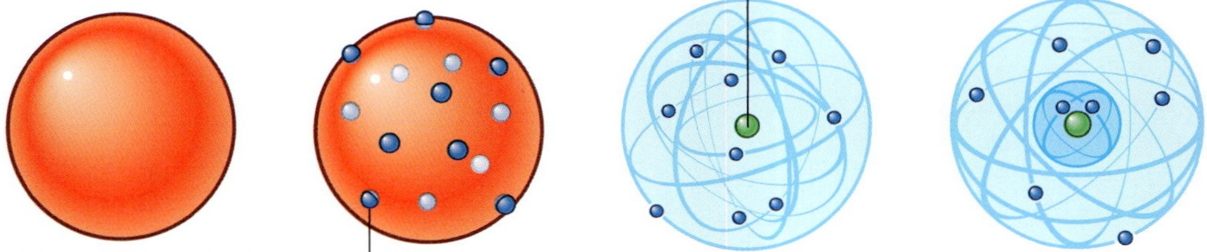

A massive star explodes as a supernova, scattering the elements in its core. They may become the building materials of new stars and their planets.

BUILDING THE ELEMENTS

The simplest atomic nucleus is that of hydrogen, a single positively charged proton. Atoms of other elements contain more protons, with uncharged neutrons to hold them together. The number of protons is called the atomic number (here in brackets). Usually the protons' positive charge is balanced by an equal number of orbiting electrons.

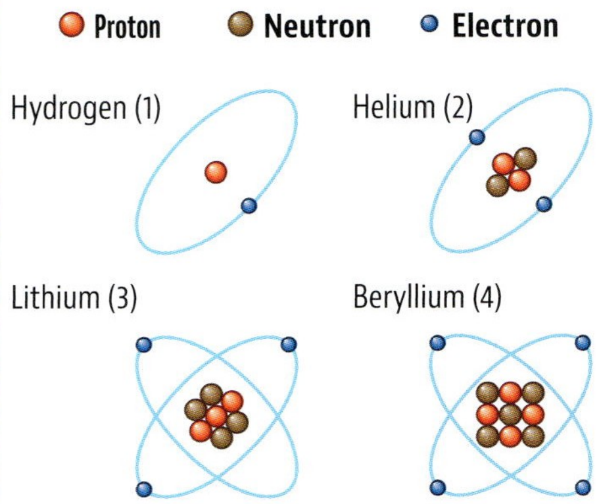

SCIENCE WORDS

Electron: A negatively charged particle found in every atom.
Neutron: A particle that is electrically neutral (uncharged) and has approximately the same mass as the proton. Outside the nucleus it decays into a proton, an electron, and an antineutrino.
Proton: A positively charged particle with approximately the same mass as the neutron. Protons are found in the nuclei of all atoms.

an electric charge equal but opposite to that of the electron. The mass of the proton is nearly 2,000 times as great as that of the electron. The other type of nucleon is electrically neutral (uncharged) and so is called the neutron. Its mass is approximately equal to that of the proton. Protons, neutrons, and electrons were built up into atoms in the first few minutes of the Universe, in the fireball of what is known as the Big Bang some 15 billion years ago. However, only the simplest and lightest atoms were formed then.

Since that time, light atoms have been welded together in the centers of stars to make more complex atoms. That is where the carbon, oxygen, nitrogen, and other atoms in the human body originated. If a star explodes as a supernova, these and heavier atoms are spread through interstellar space.

SCIENCE WORDS

Atomic number: The number of protons in an element's nucleus. It equals the number of electrons in the normal atom and is the element's numerical position in the periodic table.

Isotope: Any of the varieties of a chemical element that are chemically identical to one another, but whose atoms differ in their relative atomic mass. The atoms of the isotopes of a particular element have the same number of protons in the nucleus and the same number of electrons surrounding the nucleus, but they have different numbers of neutrons in the nucleus.

STEALING ELECTRONS

The two rare isotopes of hydrogen have their own names, deuterium and tritium. They have one or two neutrons, respectively. Most isotopes, like the two carbon isotopes shown here, have no special name. They are identified by the name of the element combined with the mass number, which is the number of protons and neutrons in the nucleus.

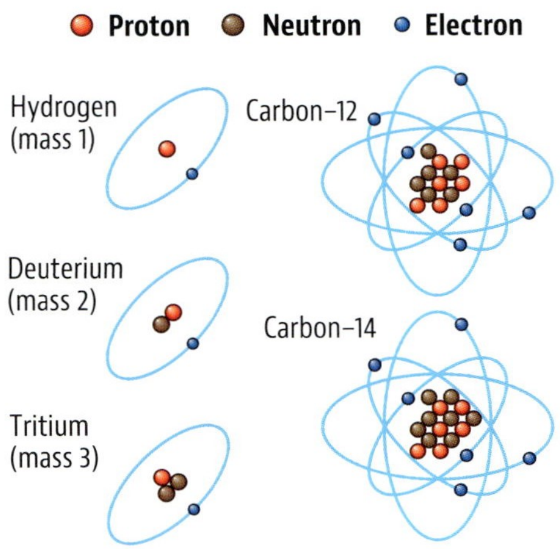

NIELS BOHR

Danish physicist Niels Bohr (1885–1962) found the first clues to the structure of the atom. He assumed the electrons circled the nucleus like planets orbiting the Sun. A major problem was that, according to the theory then existing, whirling electrons should radiate all their energy in a burst of electromagnetic radiation and fall into the nucleus in a fraction of a second. Bohr could not explain why this did not happen, but went ahead and assumed the electrons stayed in fixed orbits. He calculated the size and energy for the simplest atom, hydrogen. Bohr won the Nobel Prize in Physics in 1922. He helped to develop quantum theory, which deals with the physics of the very small, and which has replaced his simple model of the atom with a more complex one.

Atoms and Isotopes

Atoms that behave the same way when they take part in chemical reactions can turn out to be different. The reason lies in the nucleus. Although their nuclei all have the same number of protons, they may have different numbers of neutrons

There are about 90 kinds of naturally occurring, chemically different atoms on the Earth. They behave very differently in chemical reactions—some react strongly; some react scarcely at all. All these differences are a result of the activity of electrons in the outer parts of the atoms. Atoms join and separate as they gain, lose, or share electrons. The behavior of the atoms depends on the number of electrons in the atoms that are reacting, and that depends on the atom's nucleus.

In ordinary conditions, the number of electrons in an atom equals the number of protons in the nucleus, so that the atom is uncharged overall. This number is called the atomic number, and it is what defines an element. Fluorine has nine electrons and it has that number because there are nine protons in its nucleus. However, it has a strong tendency to gain an extra electron because 10 electrons make a stable system—that is, one that is not easily changed. So fluorine reacts strongly with other atoms from which it can gain an electron. It reacts especially strongly with sodium, which has 11 electrons, and which "tries" to shed one of them to leave just 10 electrons. A sodium and a fluorine atom join together, and an electron is transferred from the sodium to the fluorine, giving out energy in the form of heat.

But two atoms can have the same atomic number and yet differ in important ways. Although they have the same number of protons in the nucleus and are therefore chemically identical, they can have different numbers of neutrons. Ordinary hydrogen has a nucleus consisting of one proton and no neutrons. A rarer form has one proton joined to one neutron in the nucleus and is called deuterium. A still rarer form has one proton and two neutrons in the nucleus, and is known as tritium.

Exploring Isotopes

These different forms of an element are called isotopes of that element. They were first discovered by J. J. Thomson, who found two varieties of the gas neon. His student Francis Aston (1877–1945) studied isotopes thoroughly when he made very accurate measurements of the masses of atoms from 1919 onward.

TRY THIS

Making Heavy Water

Deuterium is the isotope of hydrogen with a neutron, as well as a proton, in its nucleus. Oxygen combines chemically with hydrogen to form water (H_2O). When oxygen combines with deuterium, the combination is called heavy water (D_2O). It is called heavy water because deuterium was originally known as heavy hydrogen. All natural water contains a tiny fraction of heavy water (0.003 percent). In this project, we will increase the amount of heavy water in a sample of ordinary water.

What to do

Nearly fill a glass jar or beaker with water. Add five teaspoonfuls of vinegar. Wrap one end of each of two wires around the terminals of a battery. Connect alligator clips on the other ends of the wires to the ends of two very thick pencil leads. Dip the leads into the watery vinegar (making sure they do not touch each other). Leave for a while.

If you look carefully, you will see small bubbles of gas on the pencil leads. On the lead connected to the positive (+) terminal of the battery the gas is oxygen. On the lead connected to the negative (–) terminal of the battery it is flammable hydrogen. The current is decomposing the water into its component elements, oxygen and hydrogen (in a process called electrolysis). There is also some heavy water in the vinegar solution, but it is not so easily broken down by electrolysis as normal water is. As you leave the experiment running, the concentration of heavy water very gradually increases. The same process has been used on a large scale for making heavy water for some kinds of nuclear reactor.

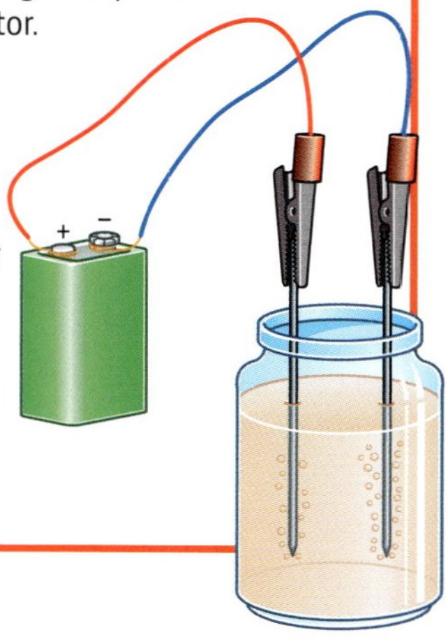

Bubbles of oxygen form around the pencil lead connected to the battery's positive terminal. The bubbles on the other pencil lead are hydrogen.

Aston formed beams of fast-moving ions. Ions are atoms that have lost or gained one or more electrons to form electrically charged particles. They can be accelerated by passing them through an electric field. Magnetic fields have the effect of bending the paths of these electrically charged particles.

Aston passed his ion beams through both electric and magnetic fields, and was able to separate them into beams depending on the charge and mass of the ions. He called his device a mass spectrometer because it spread out the beams of ions in much the same way as an optical spectrometer spreads out a beam of light.

Such experiments show that, for example, chlorine atoms are of two main types. Approximately three-fourths has a mass 35 times that of the hydrogen atom; the other fourth has a mass 37 times that of the hydrogen atom. This explained why chemists had already found that the mass of the chlorine atom is 35.45 times that of hydrogen: they had been measuring an average. The mass of an atom is expressed as its relative atomic mass, or r.a.m. The r.a.m. of the most common isotope of carbon, with six protons and six neutrons, is defined to be exactly equal to 12; on this scale the r.a.m. of hydrogen is 1.008.

TRY THIS

Isotopes and Decay

Some atoms form isotopes, which have an abnormal number of neutrons in their nuclei, and some isotopes are unstable. They become stable again by ejecting particles from their nuclei in a process known as decay. In this project, you will make models of simple isotopes and then look at a typical decay process.

What to do

Roll some balls of red modeling clay to represent protons and blue clay to represent neutrons. First, make models of the nuclei of the three isotopes of hydrogen, which are normal hydrogen (H), deuterium (D), and tritium (T). They are very simple: the normal hydrogen nucleus has just one proton, deuterium's nucleus has one proton and one neutron, and the nucleus of tritium has one proton and two neutrons. Notice that deuterium is twice the size (and weight) of normal hydrogen, and tritium is three times the size and weight.

Normal carbon has 12 particles in its nucleus (six protons and six neutrons), and is also known as carbon–12. But carbon has an isotope, called carbon–14, with 14 particles in its nucleus. There are still six protons, but there are two extra neutrons, bringing their total to eight. Make a model of this nucleus from balls of modeling clay. Carbon-14 is unstable and radioactive. It becomes more stable when one of the neutrons in its nucleus changes into a proton; that is called radioactive decay. To model this event, remove one blue ball, and replace it with a red one. Do you recognize the new nucleus you have just made? It is a nitrogen atom (with seven protons and seven neutrons).

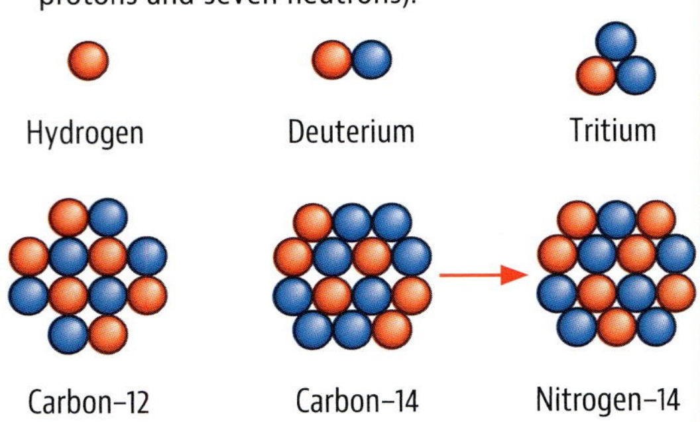

The top row shows the three isotopes of hydrogen. The first and second models beneath are the two isotopes of carbon, carbon-12 and carbon-14. Carbon-14 is radioactive and decays to form nitrogen-14, the nonradioactive form of nitrogen.

Stability and Instability

There must be at least as many neutrons as protons to make a stable nucleus—that is, one that will last unchanged. If there are too few or too many neutrons, the nucleus will disintegrate, or break down. It will do this by giving out particles or breaking in two to form other nuclei. That can happen repeatedly until stable nuclei are formed. The breaking down of a nucleus, transforming it into others, is called radioactivity. Another type of radioactivity involves just giving out energy in the form of radiation, without changing the identity of the nucleus.

In nuclei of low atomic number, there can be stable nuclei with equal numbers of neutrons and protons, as in the cases of carbon (atomic number 6), oxygen (atomic number 8), and calcium (atomic number 20). But because all protons have positive electrical charge, they tend to repel one another. As a result, heavier nuclei need

FRANCIS ASTON

Francis Aston (1877–1945) gained the first clue to the existence of isotopes when working with J. J. Thomson at Cambridge in 1913. Further study was interrupted by World War I (1914–1918). In 1919, Aston built a much improved version of the equipment he had previously used. The device revealed that atoms that were chemically identical could have slightly different masses. Nearly every element studied was found to have several isotopes. Aston announced the whole-number rule, which states that the masses of isotopes are whole-number multiples of the mass of the hydrogen atom. He also discovered that elements of odd-numbered relative atomic mass (r.a.m.) usually have isotopes of odd-numbered r.a.m., and those of even r.a.m. usually have isotopes of even r.a.m. Aston spent the rest of his career developing accurate mass spectrometers.

SCIENCE WORDS

Ion: An atom or molecule that has lost or gained one or more electrons, so gaining an electric charge.

Mass spectrograph: An instrument that separates fast-moving ions in a beam according to their mass. Similar devices are the mass spectrometer and the mass spectroscope.

Relative atomic mass: The mass of an atom expressed in atomic mass units (amu). The atomic mass unit is defined as one-twelfth of the mass of the common isotope of carbon and is approximately equal to the mass of the hydrogen atom.

THE MASS SPECTROGRAPH

The collimating slits let through a narrow beam of ions. The combination of the electric field and the magnetic field causes ions of different mass to move in curves of different radius, striking the photographic plate at different positions.

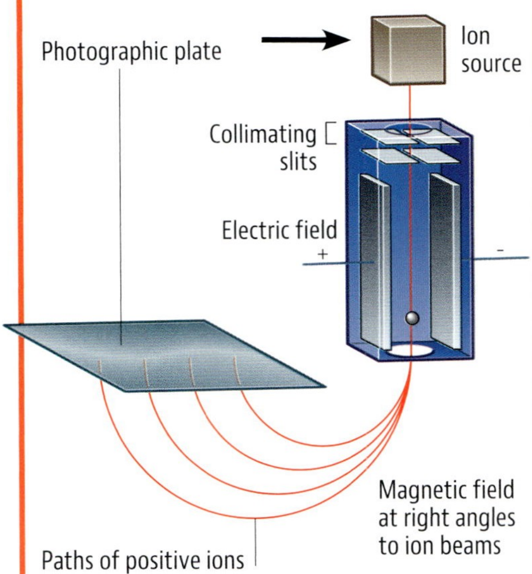

SCIENCE WORDS

Calutron: A large mass spectrometer used for separating the isotopes of uranium.
Centrifuge: A machine with a fast-spinning chamber that creates strong "artificial gravity" to separate materials of different densities. It is used in the enrichment of uranium to make nuclear fuel.
Electrolysis: The decomposition of an electrolyte—a liquid that conducts electricity—by electric current, using two electrodes.
Gas diffusion: A technology used in the enrichment of materials that are used in nuclear reactors and the making of nuclear weapons.

extra neutrons to hold the nucleus together against the force of repulsion between the many protons. The most stable isotope of uranium, atomic number 92, has 146 neutrons, but even so it is weakly radioactive.

Separating Isotopes

Separating isotopes can be very difficult, but the problem had to be solved in World War II in order to build the first nuclear bomb. Today, the same technology has countless peaceful uses. Specific isotopes are needed for industrial measurement, medical diagnosis, and biological study.

Once scientists had discovered the existence of isotopes, they had to separate them in order to study them. That was not easy to do because the isotopes of an element are chemically identical and so are affected in the same way by a given chemical process. The principle behind the mass spectrograph

This is a calutron, a kind of mass spectrometer, that was designed in the 1940s to separate the isotopes of uranium. Nuclear physicists used it to produce quantities of enriched uranium-235.

TWO WAYS OF SEPARATING ISOTOPES

A centrifugal separation unit receives a mixture of gaseous isotopes as "feedstock" and returns one stream of gas that is enriched in the lighter isotope and another that is depleted.

At each stage of a gas-diffusion separation plant the depleted part of the output is sent on, while the enriched part is passed back to an earlier stage.

CENTRIFUGAL SEPARATION

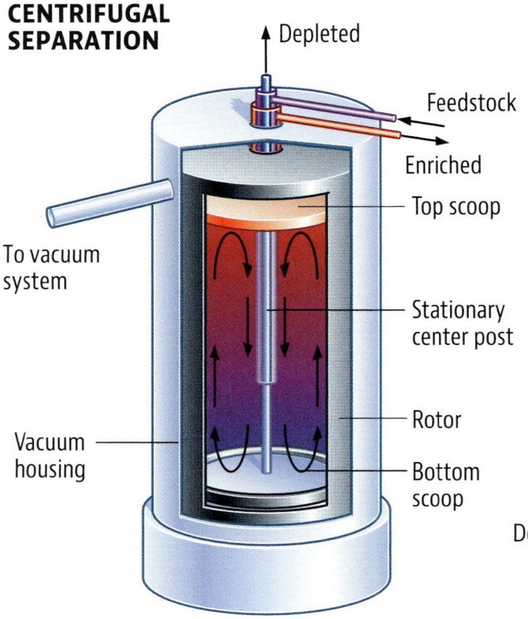

In centrifugal separation, a gaseous mixture of isotopes is passed into a fast-spinning chamber. The heavier isotopes are thrown outward to the edge of the rotating chamber slightly more strongly than are the lighter isotopes, and so there is partial separation. Streams of gas are collected from the center of the centrifuge and the edge of the centrifuge by two "scoops."

GAS-DIFFUSION SEPARATION

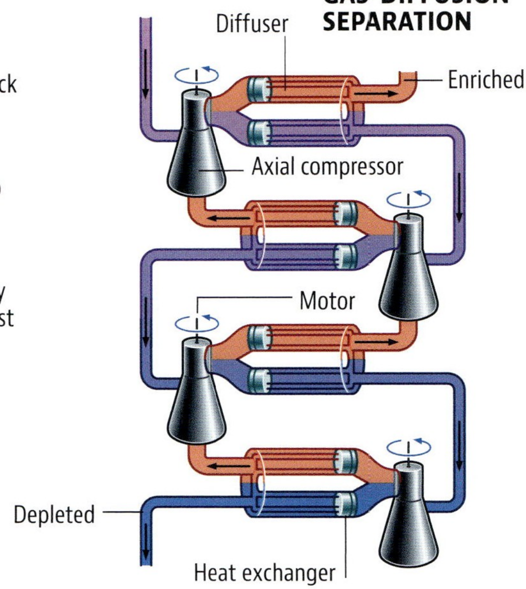

In gas-diffusion separation, uranium is combined with fluorine to make uranium hexafluoride gas. It is allowed to diffuse through a porous membrane. The uranium-235 moves slightly faster, so the gas emerging is slightly enriched. The gas is put through the same process thousands of times, gradually increasing the proportion of uranium-235. Then electromagnetic separation is used for the final enrichment.

was developed for large-scale processes. Substances containing a mixture of isotopes are ionized, and a beam of the ions is passed through electric and magnetic fields. The beam splits up because the ions that have less mass are more readily deflected.

The first to succeed in separating isotopes on a larger scale was the American chemist Harold Urey (1893–1981) who, in 1932, separated the rare isotopes of hydrogen, called deuterium and tritium, from the common isotope. He did it by the electrolysis of water—that is, by breaking down water into hydrogen and oxygen by passing an electric current through it. The hydrogen released bubbles off, but the heavier isotopes do so more slowly. The water left behind is slightly enriched in the rare isotopes.

MASS AND WEIGHT

The mass of an object remains the same wherever it is on the Earth. It even stays the same if it is sent to the Moon or launched by rocket into outer space. However, an object's weight can change depending on the local force of gravity

Mass is a measure of the amount of matter in an object. That is why an object's mass always remains the same, wherever it is. But the weight of an object is the force acting on it by the gravitational attraction of the Earth (or any other nearby planetary body). As a result, an object's weight depends on its distance from the Earth. It is very slightly less at the top of a high mountain than at sea level. On the surface of the Moon the same object would weigh only about one-sixth of its weight on the Earth. That is because the Moon's force of gravity is only one-sixth of the Earth's.

The scientific unit of mass is the kilogram (kg); the unit of weight is the newton (N). An object's weight is equal to its mass multiplied by the acceleration due to gravity (which is also called the acceleration of free fall). Because this equals 9.8 meters per second per second (9.8 m/s^2) at the surface of the Earth, there is a simple relationship between weight (in newtons) and mass (in kilograms): weight = 9.8 * mass.

Therefore a person whose mass is 50 kg weighs 490 newtons. The same person would weigh only 82 N on the Moon—and a huge 1,294 N on the giant planet Jupiter!

TRY THIS

Do Heavy Objects Fall Faster than Light Objects?

Which falls faster—an iron cannonball or a feather? Find out the answer using a coin and a paper disk.

What to do

Cut a disk of paper very slightly smaller than a coin. With your arms held out sideways and hands palms down, hold the disk horizontally with the fingers of one hand around its edge. Do the same with the coin in the other hand. Let go of the disk and the coin at the same time. Which hits the ground first?

The paper falls more slowly, but not because it is lighter. It is slowed by air resistance and "floats" on the air as it falls. If there was no air, the two objects would fall at the same rate. A NASA astronaut proved this by dropping a hammer and a feather onto the airless surface of the Moon. Both objects hit the ground at the same time.

Now put the paper on top of the coin and, holding just the coin by its edges, without touching the paper, drop the two together. This time the paper travels with the coin and hits the ground at the same time. Air pressure above the coin keeps the paper in place, so that the paper hitches a ride with the falling coin.

Repeat the experiment using a book and a rectangle of paper cut slightly smaller than the book. Lay the paper flat on the book's cover, and drop them both together.

Drop the coin and the paper disk at exactly the same time.

NASA's Psyche spacecraft atop a SpaceX Falcon Heavy rocket lifts off from Kennedy Space Center's Launch Complex, Florida, in 2023. Powerful motors lift rockets into space against the Earth's gravity.

SCIENCE WORDS

Acceleration: The rate of change in a moving object's velocity. It is a vector quantity.
newton (N): The derived SI unit of force. It is the force required to give a mass of 1 kilogram an acceleration of 1 meter per second per second.
Weight: The force with which a mass is drawn toward the Earth (by the force of gravity).

Everyday Weight

Scientists are always very careful to distinguish between mass and weight, but this difference is not so important in everyday life. In fact, for ordinary measurements mass units are generally used to express weights. A person is described as weighing 50 kg, and might buy 5 kg of potatoes.

Sometimes it is necessary to convert from one system of mass/weight to another. To convert from kilograms to pounds, multiply by 2.2 (thus 50 kg = 110 lb). To convert from pounds to kilograms, divide by 2.2.

SLOWING DOWN GRAVITY

To study the speed of a falling ball, the Italian scientist Galileo Galilei slowed it down by rolling it down a slope.

He measured how far the ball rolled in equal time intervals, and found that speed increased uniformly with time.

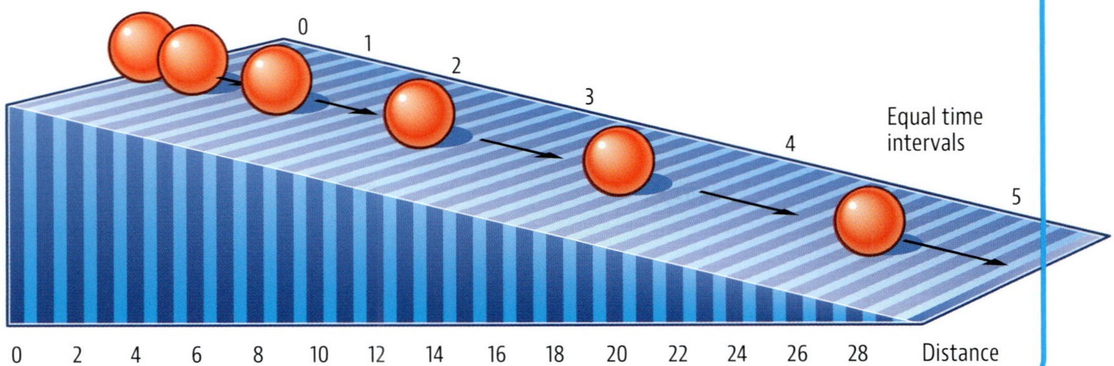

SCIENCE WORDS

Center of gravity: Also called the center of mass, the point at which an object's total mass appears to be concentrated and at which it acts.
Force: An influence that changes the shape, position, or movement of an object.

A weightlifter strains to lift a heavy weight against the force of the Earth's gravity. He would find it much easier to lift the weight if he tried it on the Moon.

Machines for Weighing

The earliest weighing machines were balances. They are still used throughout science and industry. Balances consist of a horizontal beam pivoted at its center so that it balances. A pan hangs from each end of the beam. To find the weight of a particular quantity, it is put in one pan, and known weights are added to the other pan until it balances. To weigh out a particular quantity, for example a kilogram of rice, a mass of 1 kilogram is put on one pan. Rice is then poured onto the other pan until the beam is again horizontal. Of course, such a "weighing" exercise is not really finding out what something actually weighs. It is manipulating masses. This type of balance would work just as well on the Moon because it is a method of comparing masses. A spring balance is different, however. Most mechanical

GRAVITY AND LIGHT

A gravitational field can bend a beam of light. When light from a distant object passes on either side of a galaxy, the galaxy's powerful gravity bends the light rays. An observer on Earth sees two images of the same object, one on each side of the galaxy.
(This diagram is not to scale.)

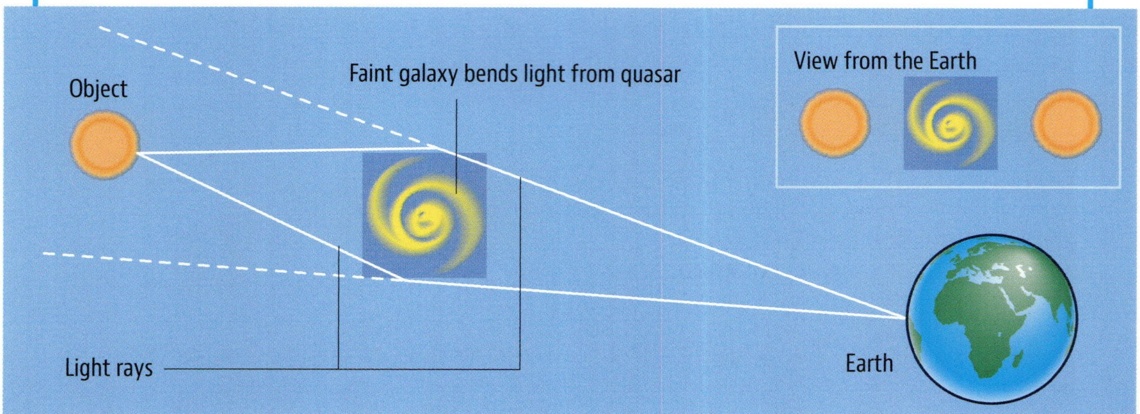

kitchen scales are spring balances, with a vertical coiled spring and a pan on top that compresses the spring. A pointer that is worked by the movement of the spring indicates weights on a marked dial. This device measures the effect of the force of gravity on an object's mass. It is therefore a force meter; similar devices called newton-meters are used in physics laboratories. The spring balance measures weight and would register only a sixth of an object's "Earth weight" if the same object were weighed in the same way on the Moon.

TORSION BALANCE

The English physicist Henry Cavendish (1731–1810) made this apparatus in 1798. A gravitational force attracted two small balls toward two larger ones, causing twisting in the suspension. From the amount of twisting, Cavendish calculated the value of the gravitational constant.

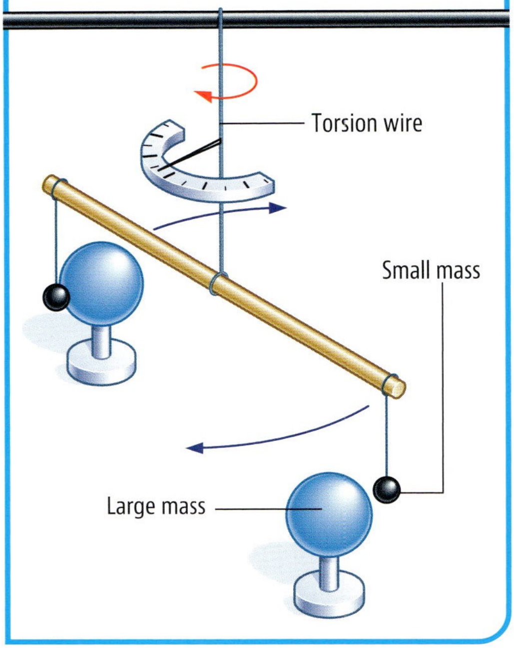

Force of Gravity

Every object is attracted toward every other object by a force known as gravitation, which arises because objects have mass. The force of gravity is simply the force of attraction between an object and the much more massive Earth. When you drop something, it is this force that makes it fall to the ground. The actual size of the force between any two objects depends on their masses and their distance apart. In mathematical terms, the force is proportional to the product of the masses divided by the square of the distance between them. As a result, the closer they are together, the stronger the force of attraction. If the masses are m_1 and m_2, and the distance between them is d, the force F between them can be expressed as

$$F = G \frac{m_1 * m_2}{d^2}$$

where G is the gravitational constant. This type of equation is called an inverse square law; the strength of a quantity (here, force) lessens with the square of the distance from a particular point.

GALILEO

Galileo Galilei (1564–1642) was an astronomer and physicist who was born in Pisa, Italy. One of his main subjects of study was gravity. Observing the swinging lamps in Pisa cathedral, he realized that the regular swinging motion of a pendulum might be used to regulate a clock. He studied falling objects by dropping weights from towers and rolling balls down inclined planes. Galileo also made one of the first telescopes, which he used to discover craters on the Moon, sunspots, and the four moons of Jupiter. He also backed Copernicus (1473–1543) by saying that the Earth orbits the Sun, and not the other way around, as was then believed.

FALLING OBJECTS

When an object falls under the force of gravity, does it fall at a constant speed, or does it get faster and faster? In other words, does it accelerate? This question puzzled early scientists until Galileo Galilei carried out some experiments to find the answer.

Galileo's experiment illustrated on page 21 shows how he found that the speed of a ball rolling down a slope continues to increase—in other words, it accelerates. Galileo also found that every freely falling object has the same acceleration. Now called the acceleration due to gravity, it has the value 32 feet per second squared, or 32 ft/s^2 (9.8 m/s^2). Galileo is also said to have tried to measure the acceleration of a cannonball if it were dropped from the top of the Leaning Tower of Pisa. The diagram on the facing page shows what the results would have been if Galileo had been able to make such calculations. The cannonball would have reached a speed of 32 ft/s (9.8 m/s) after 1 second, a speed of 64 ft/s (19.6 m/s) after 2 seconds, and so on. The speed increases, but the acceleration is unchanging. In practice, an object falling through air does not keep getting faster and faster. The air resists the downward movement of the object, and this air resistance acts as an upward force called drag. The drag increases as the object's speed increases so that eventually the object can go no faster.

SCIENCE WORDS

Air resistance: A force, also called drag or wind resistance, that resists the movement of an object through the air. It is overcome by streamlining.
Terminal speed: The maximum speed at which an object falls under the influence of gravity.

Terminal Speed

Most falling objects reach a constant terminal speed of about 177 ft/s (120 mph; 54 m/s). Think of the damage hailstones could do if they fell any faster! Drag is greater on an object with a large area than it is on one with a small area. This is the principle of a parachute, which when open gives a falling human a terminal speed of nearly 21 ft/s (14 mph; 6.3 m/s). That is the speed at which

Skydivers reach their terminal speed, about 177 ft/s (54 m/s), after about 12 seconds. They fall at this constant speed until they open their parachutes and rapidly decelerate, reaching the ground at about 21 ft/s (6.3 m/s).

ACCELERATION DUE TO GRAVITY

The figures below show how an object would fall in the first three seconds after being dropped from the top of the Leaning Tower of Pisa. Its speed increases, but its acceleration is constant at 32 ft/s^2 (9.8 m/s^2).

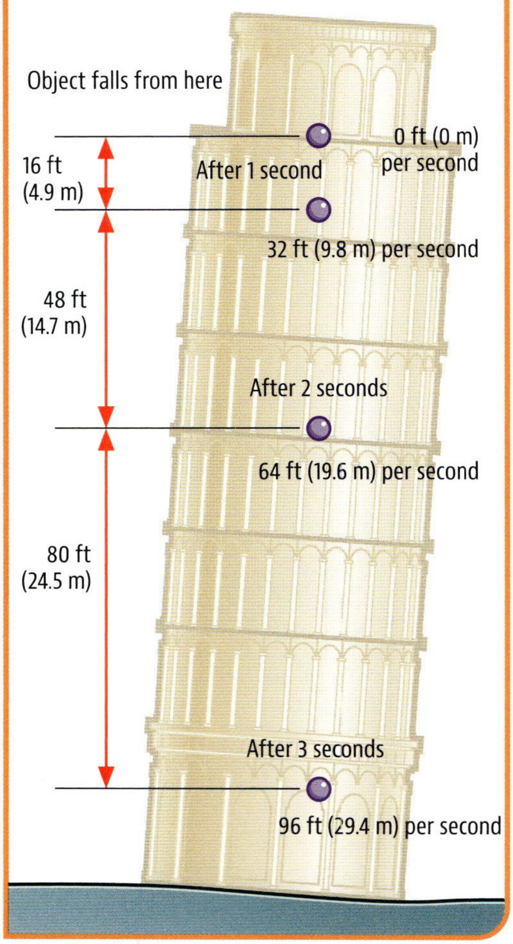

a parachutist hits the ground. The principle can be tried using a sheet of paper. Drop it, and it flutters to the ground because of the high drag effect acting on it. If the paper is then crumpled into a tight ball it will fall much quicker because there is much less drag.

An object falling under the force of gravity is said to be in free fall (and acceleration due to gravity is also known as acceleration of free fall). It has the symbol g, and crops up in many physical formulas to do with mechanics, such as the formula for calculating the time of a pendulum's swing and equations for calculating pressures under water.

VECTORS AND SCALARS

Most quantities in physics are expressed as a number and some unit, such as 25 kg or 110 volts. These last examples are called scalars. But what is the difference between a speed of 50 km/h and a velocity of 50 km/h in the direction of Chicago? The first (speed) is a scalar, but the second (velocity) is a vector quantity.

A vector quantity always has its direction specified, whereas a scalar is a pure number. A given quantity can be either, and sometimes the difference is very important. If you told some shipwrecked sailors on a life raft that there was an island only 3 miles (4.8 km) away, they might be relieved. But it would be much more useful for them to know that there was

SCIENCE WORDS

Scalar: A quantity that has magnitude but (unlike a vector) no specified direction. Examples of scalars are speed and mass.

Triangle of vectors: A way of adding vectors. The first vector is drawn as a line at the correct angle, with the length of the line representing its magnitude. The second vector is drawn from the end of the first line, again at the correct angle and of the correct length. A third line joining the beginning of the first line to the end of the second line (completing the triangle) gives the magnitude and direction of the sum of the vectors.

Air-traffic controllers use vectors to indicate the position of airplanes. They need to know how far away the airplanes are and in what direction. Vectors are plotted on a screens.

an island 3 m (4.8 km) away to the north. They would then know which way to paddle the life raft. The first piece of information, "3 miles (4.8 km) away" is a scalar quantity; "3 m (4.8 km) away to the north" is a vector. Several quantities in physics are vectors. They include velocity, acceleration, and most forces.

Adding Vectors

Adding scalar quantities is easy. A piece of string 3 ft (0.9 m) long added to a piece 4 ft (1.2 m) long gives a total length of 7 ft (2 m) of string (ignoring the string used to tie the knot). Adding vectors is trickier. If a cart is pushed 30 ft (9 m) to the east and then 40 ft (12 m) to the south, where does the cart end up? The answer is 50 ft (15 m) away from where you started, in a more or less south-easterly direction. Notice that the cart does not finish up 70 ft (21 m) away, which is the result you would get if you merely added the two distances. The math is more complicated this time.

One way of adding vectors is to draw a plan. Make the length of a line stand for the size of the vector, and draw it in the correct direction. From the end of the first line draw a second line in the direction of the second vector, again with length standing for size. A line drawn between the beginning of the first line and the end of the second one represents the sum of the vectors and its direction. The construction you have made is called a triangle of vectors.

TRIANGLES AND PARALLELOGRAMS

The triangle of vectors (right) shows how to add vectors, in this case two forces. The effect of combining a force F_1 in one direction and another force F_2 in a different direction (the lengths of the arrows indicate the sizes of the forces) is called the resultant. The third side of the triangle shows the resultant's size and direction.

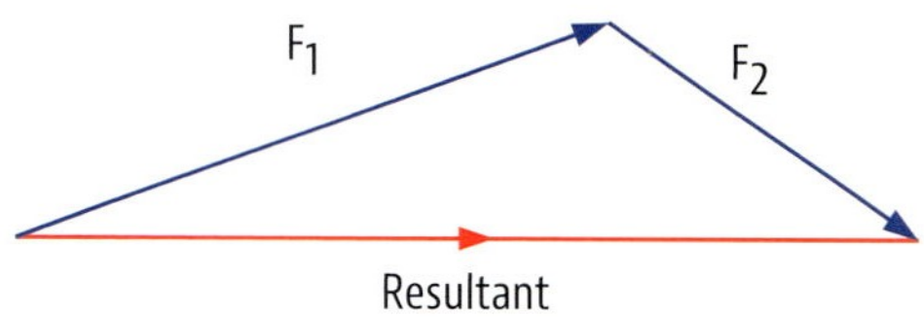

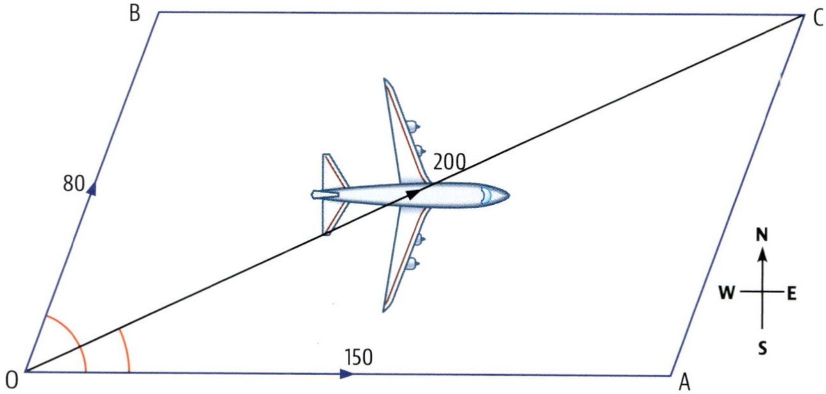

Problem: An airplane navigator wants to fly a plane 200 miles (322 km) from O to C. But there is a northerly wind along the line from O to B. What course should the navigator set? The correct course is due east, toward A. The parallelogram of vectors then shows that the plane will actually fly along the line from O to C, as required.

FORCE AND ACCELERATION

Force is what makes an object move or stop. The ease with which something moves depends on its mass. A moving object has a certain speed, and if its speed changes it accelerates. Force, mass, and acceleration are interconnected.

Force can make things move, stop moving, move faster or slower, or change the direction in which they move. It can also make things change shape, perhaps by squeezing them or stretching them. If a rubber band is stretched or a pencil is snapped this uses force. Even a humble paper clip exerts a force when it holds two pieces of paper together.

All the forces mentioned so far have to do with objects in contact with one another. But there are other forces that act at a distance. Gravity is a force that pulls things down to the surface of the Earth. A magnet exerts a force when it picks up an iron nail. There are also forces that act between electric charges. Nearly all of physics is concerned with forces of one sort or another, but they are most obvious when they have to do with movement, with what scientists call motion.

When the hockey player hits the puck, it slides across the ice and, according to Isaac Newton's laws, would go on forever if the force of friction did not slow it down.

ISAAC NEWTON

Sir Isaac Newton (1642–1727) was an English mathematician, astronomer, and physicist. He was born in Lincolnshire and in 1661 studied at Cambridge University. In 1669, he became professor of mathematics there. In math, Newton developed calculus, a system that deals with changing quantities. In astronomy, he made one of the first reflecting telescopes and worked out how the Moon orbits the Earth using the law of gravitation,a major contribution to physics. He studied light and produced the Sun's spectrum by passing sunlight through a glass prism. He also formulated his three laws of motion. In 1705, he was the first scientist to be knighted. He is buried in Westminster Abbey in London.

The Laws of Motion

If a ball is rolled along the ground, it does not keep on rolling forever. It soon slows down and stops. But have you ever wondered why? In the 17th century, the English scientist Isaac Newton (1642–1727) asked himself this question and came up with a set of rules that apply to all moving objects. Together, these rules are now known as Newton's laws of motion.

A person kicking a ball demonstrates the third law of motion (see page 30). The player's foot applies a force to the ball, and the ball applies an equal force to the player's foot.

The first law states that an object at rest will stay at rest, or a moving object will go on moving, unless a force acts on it. According to Newton, if our ball started rolling because it was given a push, force has been applied. It then stopped rolling because it was acted on by another force. In this case, the second force was friction between the ball and the ground. That is why a ball will roll farther on a smooth surface than it will on a rough one—try rolling a marble across a wooden floor (low friction), then across a carpet (high friction).

Newton's second law of motion involves acceleration, which is the rate at which a moving object changes speed. The law says that the force acting on an object is equal to its mass multiplied by its acceleration. If something is pushed (force is applied), it will move off at a certain speed. However, to make it move faster and faster, force has to keep on being applied. A spacecraft returning to the Earth from the Moon travels at a fairly constant speed until it gets close to the Earth. Then the Earth's gravity (a force) has more effect and makes the spacecraft accelerate. Strictly speaking, the term velocity should be used here instead of speed (velocity is speed in a specified direction).

TRY THIS

Rocket Balloon

The gases under pressure inside a rocket push in all directions, but the only way they can escape is through the nozzle at the rear of the rocket. A reaction force pushes in the opposite direction, and it is this force that makes the rocket move forward. In this project, you use reaction to make a balloon into a rocket.

What to do

Cut a drinking straw in half (throw away one half). Thread one end of a piece of string about 12 ft (3.5 m) long through the straw. Place two sturdy chairs about 8 ft (2.5 m) apart, and tie the ends of the string to the chair backs. Move the chairs apart until the string is taut. Cut two short lengths of masking tape, and blow up the balloon. Twist the neck of the balloon and hold it to keep in the air. This is where you may need some help. Slide the straw along the string and tape it onto the inflated balloon. Then let go of the neck of the balloon—it will jet along the string under rocket power.

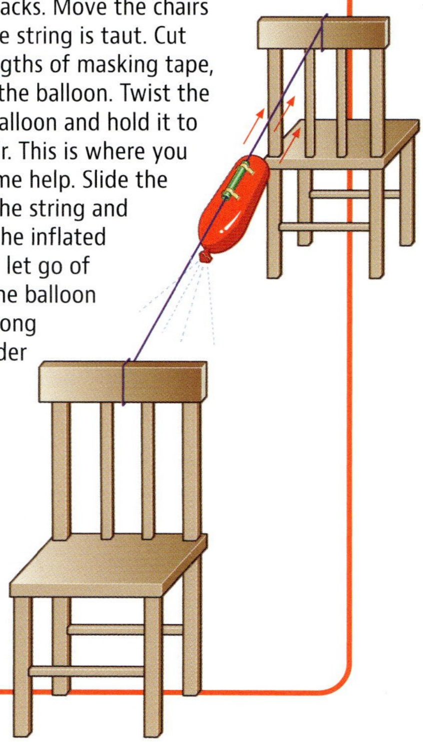

The escape of compressed air powers the rocket balloon along its string.

The third law of motion concerns two objects. It states that when one object exerts a force on another, the second object then exerts the same force on the first, but in the opposite direction. The first force is an action and the second a reaction, and the law is sometimes stated as "action and reaction are equal and opposite."

When a book is dropped, gravity is the action force that makes it fall. There is also an equal reaction force between the book and the Earth, but it is undetectable because the mass of the Earth is huge compared with that of the book. The forces between the Earth and the Moon are easier to grasp. The Earth's gravity pulls on the Moon and keeps it in orbit. The Moon's gravity pulls on the water in the Earth's oceans, causing the daily tides.

Another example of the third law is the principle of the rocket. The hot burning gases in a rocket expand and push in all directions. Those that push on the closed front end of the rocket are balanced by a reaction, which acts in the opposite direction and propels the rocket along. For this reason, a rocket is known technically as a type of reaction motor.

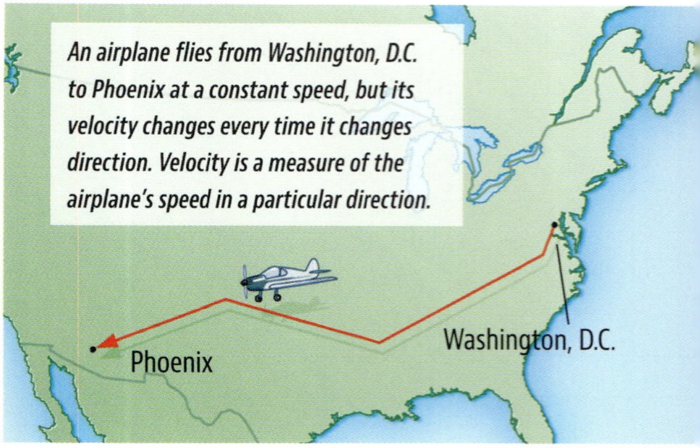

An airplane flies from Washington, D.C. to Phoenix at a constant speed, but its velocity changes every time it changes direction. Velocity is a measure of the airplane's speed in a particular direction.

Inertia and Momentum

A heavy object is more difficult to get moving than a lighter one. That is because of its mass

STARTING AND STOPPING

When a drag racing car starts **(a)**, the forward push of the car's motor has to overcome friction. During the run **(b)** the car continues to accelerate as the motor continues to provide forward force, although there is some wind resistance. At the end of the run **(c)** the driver turns off the motor and releases a parachute to provide even more wind resistance and slow the car down until it comes to a standstill.

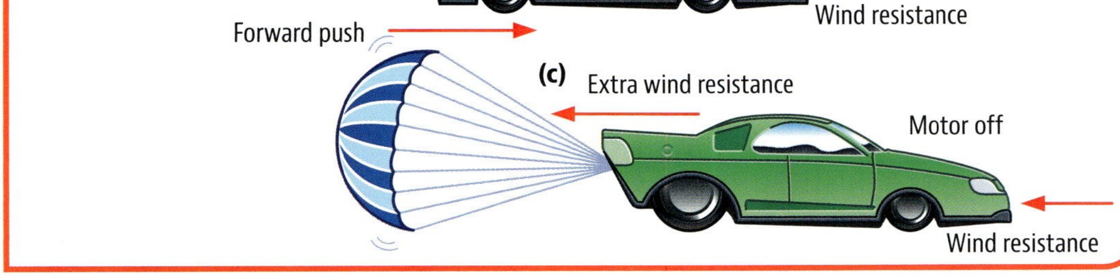

or inertia. Inertia can be thought of as an object's reluctance to move. If you are traveling in an automobile and the driver brakes suddenly, it is your inertia that makes you keep moving unless held back by a seat belt.

Once an object is moving, it has momentum, which is equal to its mass multiplied by its velocity. The more massive a moving object is, or the faster it moves, the greater its momentum. It is easy to demonstrate the difference between inertia and momentum. If you carefully place a brick on your foot, it is difficult to raise your toes—that's inertia. But if you dropped the brick on your foot, it would do a lot of damage—that's momentum! A small object moving very fast can have more momentum than a massive object moving slowly. For example, in theory a bullet fired from a Magnum revolver could have enough momentum to stop a moving automobile because the bullet's momentum is greater than that of the car.

Newton's second and third laws of motion predict that when two objects bump into each other, their total momentum after impact is the same as it was before impact. This statement is often called the principle of conservation of momentum. It explains various things in everyday life, particularly in sports. People playing pool or hockey make unconscious use of the conservation of momentum when they strike a ball or the puck. Like velocity, acceleration is a vector quantity. It is specified by a number (how large it is) and a direction, such as "5 feet per second northward." Acceleration is measured in units such as meters per second per second. Speed is a scalar quantity. It is stated just as a number and unit with no direction, such as "500 mph."

SCIENCE WORDS

Inertia: The property of an object that makes it resist being moved or, if moving, to resist a change in direction. It is a consequence of the first of Newton's laws of motion.
Momentum: The mass of an object multiplied by its velocity in a straight line.

TRY THIS

Check That Checker

This project relies on inertia. The idea is to remove the bottom checker from a stack of checkers without touching it or knocking over the stack.

What to do

Stack nine checkers in a neat pile on a shiny surface. Place the tenth checker about 1 in (2.5 cm) away from the bottom of the stack. With your finger and thumb flick the single checker hard against the base of the stack. The bottom checker should shoot off sideways, leaving the rest of the stack in place. Some people will find it easier to use the edge of a ruler to flick the bottom checker away.

Inertia is the reluctance of a stationary object to move. It is inertia that keeps the other checkers from moving even when the bottom one is forced violently sideways. If you master the trick of using a ruler to strike the checker, you can knock out any of the checkers in the stack—not just the bottom one—without disturbing the others.

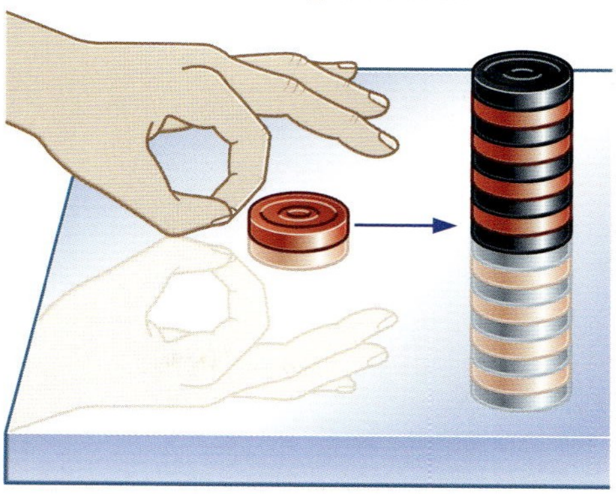

Flick the single checker hard at the base of the pile (or hit the last checker in the pile with the edge of a ruler).

MOTION IN A CIRCLE

So far we have considered mainly objects moving in straight lines and the various forces that can act on them. Slightly different rules apply if an object is moving in a curved path, particularly if it is moving around and around in a circle.

People on this type of carnival ride feel as though they are being flung outward. In fact, their velocity is constantly changing as they go around in a circle.

We are going to begin with a difficult idea. When an object moves in a straight line, the only way it can accelerate (or decelerate) is by changing its speed—by going faster (or slower). But we know from pages 26 to 27 that velocity is a vector quantity. If an object is moving at a certain speed in a certain direction (that is, if it has a certain velocity) it can be acted on by a force—for example, by giving it a sideways push—that changes its direction without changing its speed. As a result, its velocity changes. That amounts to saying that it undergoes acceleration, which is defined as change in velocity divided by time. Although it might seem unlikely at first, it is possible for something to be accelerating without its speed changing.

Think of a stone tied to the end of a piece of string and whirled around your head in a horizontal circle. Its speed is constant; but because its direction is continuously changing, it is always accelerating. The force that accelerates the stone is the pull in the string. Its direction is at right angles to the direction of the stone at any one time, directed inward toward the center of the circle. Physicists call it centripetal force.

FEEDBACK CONTROL

A mechanism for controlling the speed of an engine, particularly in steam engines, is called a centrifugal governor. This makes use of circular motion. The vertical shaft is turned by a belt **(1)** driven by the machine. As it turns, the weights **(2)** attached to the top of the shaft go around and rise **(3)**. This action raises the rod **(4)**, which reduces the power supply to the engine and slows it down. As it slows, the weights fall and so increase the power to the engine.

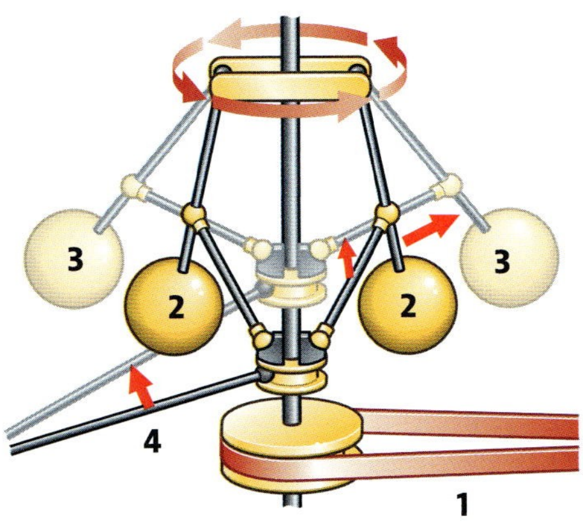

SCIENCE WORDS

Centripetal Force: The force that acts inward to keep an object moving in a circle. Centrifugal force is a force that causes an object moving in a circular path to move out and away from the center of its path, in opposition to centripetal force.

Moons and Satellites

Isaac Newton (1642–1727) worked out that a force is involved in keeping an object moving in a circle around another object. So what keeps the Moon orbiting the Earth? It is the force of gravity between the Earth and the Moon. In a way, the Moon is always falling in a circle toward the Earth. But it keeps falling past and going around again—or at least it has for the last 4½ billion years or so.

Artificial satellites orbiting any planet act in the same way under the gravitational attraction of the body that they orbit. To lift an orbiting satellite into a higher orbit, it has to be given more speed (by firing its rocket motors). When a satellite slows down, perhaps because of friction between it and the outer layers of the Earth's atmosphere, it can no longer stay in its orbit and soon spirals down toward Earth.

HAMMER THROWER

Centripetal force in the chain attached to the hammer produces its acceleration in a circle. When the thrower lets go, the hammer flies off in the direction it was traveling at the instant it was released.

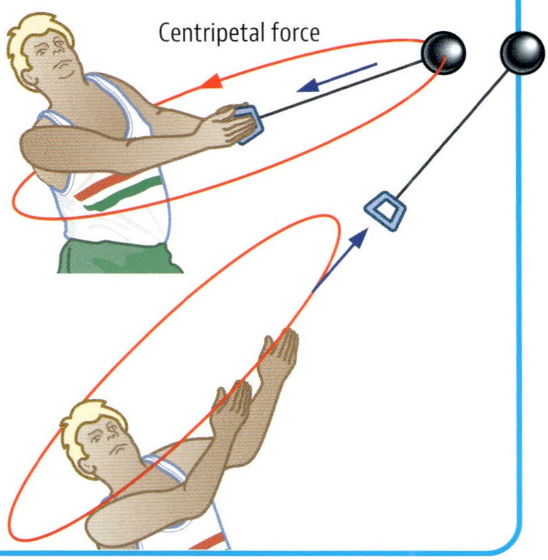

TRY THIS

Spinning Water

When objects spin around fast in a circle, there is a force trying to push them in toward the center. It is called centripetal force and is the principle behind a spin drier. The drier makes the wet laundry go around in a circle, but the water moves outward through the holes in the spinning drum, like a hammer thrower's hammer when it is thrown. This project demonstrates centripetal force.

What to do

Pour water into a large mixing bowl until the water is about 4 in (10 cm) deep. Float a small bowl on the water, then pour a little water into the bowl until the water is about 0.5 in (1 cm) deep. Place the brush of a dish brush in the bowl and use it to spin the bowl around (or use your index finger if you have no dish brush). Keep spinning until the bowl is going round really fast.

The water in the spinning bowl will move toward the edge of the bowl and "crawl" up the sides until the bottom of the bowl is dry. As the bowl slows down, the water will sink back toward the center and again cover the bottom. The water has behaved like the water in a spin drier.

Spin the small bowl around and around, and watch the water in it rise.

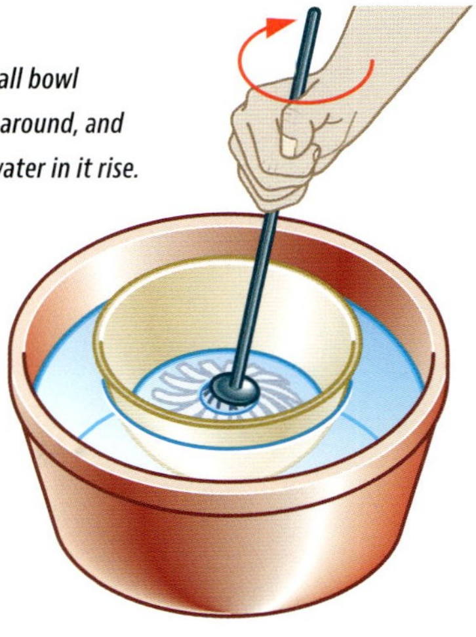

THE SWINGING PENDULUM

In 1851, at the Panthéon in Paris, the French physicist Léon Foucault (1819–1868) conducted a famous experiment using a pendulum to demonstrate the rotation of Earth.

A pendulum swings through an arc of a circle, with the suspension point at the center of the circle. Two quantities that can be varied are the length of the pendulum and the mass of the pendulum bob. Only one of them affects the timing of the swing.

There is a traditional story about the first person to observe the perfectly regular swing of a pendulum. In 1602, while the Italian scientist Galileo Galilei (1564–1642) was sitting in Pisa cathedral, he noticed the ceiling lamps swinging in the breeze. He did not have a watch, so he timed the swings by feeling the pulse in his wrist.

Galileo found that each lamp swung at a regular pace, and that lamps on long chains swung more slowly than on short ones. He also wondered whether heavy lamps swung more slowly than lighter ones. As he could not weigh the lamps, he carried out some experiments using pendulums.

LENGTH AND MASS

Galileo found that changing the mass of the bob (the weight at the end of the cord) had no effect on the time of the swing as long as the pendulum swung through only small angles.

Changing the length of the cord did change the time of the swing. This time period was halved when he shortened the cord to one-fourth of its former length.

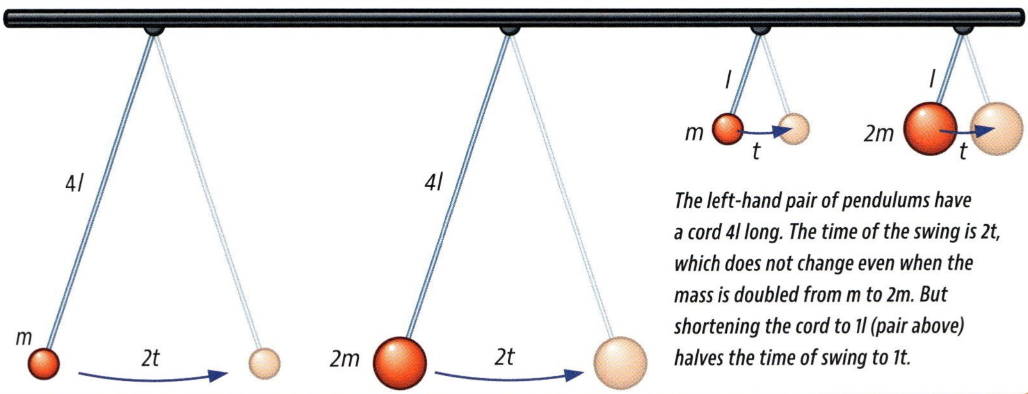

The left-hand pair of pendulums have a cord 4l long. The time of the swing is 2t, which does not change even when the mass is doubled from m to 2m. But shortening the cord to 1l (pair above) halves the time of swing to 1t.

Enter Gravity

Galileo found that the time for a pendulum's complete swing—that is, from one side, across to the other side, and back again—is proportional to the square root of the pendulum's length. The actual relationship is given by the equation:

$t = 2\pi \sqrt{l/g}$

where t is the time and l is the length of the cord. Notice that the equation has two other symbols. One is π (pi), which often turns up when circles are involved, and the other symbol is g, the acceleration due to gravity.

Uses of Pendulums

The first important application of the pendulum was to regulate the mechanism of a clock. Galileo suggested this, but the Dutch scientist Christiaan Huygens (1629–1695) built the first practical pendulum clocks in 1656.

The swinging pendulum was regulated by an escapement, which was a wheel with pointed teeth. As the wheel rotated, the teeth gave the pendulum a small push with each swing, allowing the clock's other wheels to advance a fixed amount with each swing, so moving the clock's hands forward.

TRY THIS

Swinging Low

A weight swinging at the end of a thread or string is called a pendulum. This project looks at what happens if you change the size of the weight, and what happens if you change the length of the string.

What to do

Cut a length of string to the height of a table top above the floor. Place two rulers together, trap the string between them close to one end of the rulers, and tape the rulers together at both ends. The idea is to be able to vary the length of the string by pulling it between the two rulers.

Place the rulers so that the string end overhangs the edge of the table. Hold them in place using a heavy book. You can now tie various weights to the lower end of the string to make pendulums.

First, try a small washer. Tie it to the string, pull it to one side, and then let go. Count the number of swings it makes in 10 seconds. It is easier if you have a helper to shout out when 10 seconds are up while you count the swings. Make a note of the answer. Repeat with heavier weights, and again note how many swings occur in 10 seconds.

Choose another weight, say an eraser, and tie it to the string. This time vary the length of the string by pulling it through the rulers, again counting the number of swings in 10 seconds for each length of pendulum. Make a note of your findings. You should find that the number of swings does not change when you vary the weight at the end of the string. However, the number should vary when you change the length of the string. In fact, if you shorten the string to one-fourth of its original length, the pendulum will swing twice as fast.

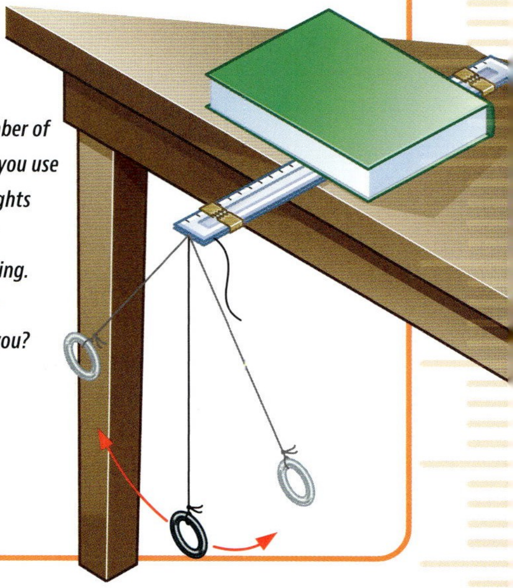

Time the number of swings when you use different weights and different lengths of string. What do your findings tell you?

ENERGY, WORK, AND POWER

In scientific terms, energy, work, and power are sometimes confused. Energy is the ability to do work, which results when a force acts over a distance, and power is the rate of doing work. There are many different kinds of energy.

When asked to do a chore at the end of a tiring day, we may say, "I can't, I don't have the energy." This is a fairly accurate statement, scientifically speaking. Energy is something that, possessed by something else, enables it to do work. As we will see, there are various kinds of energy. It cannot be made or destroyed, which is the principle of conservation of energy, but we can use energy, and in doing so energy can be changed from one of its forms to another form. The energy humans use comes from food. Food is essentially a mixture of chemicals. The processes of digestion change them into other chemicals, such as the high-energy sugar glucose. When we do work, our muscles use up glucose to provide energy.

Types of Energy

Potential energy is energy that something has because of its position. A book on a shelf, for example, has energy stored in it as potential energy. It may look the same as a book on the floor; but if it is knocked off the shelf, the falling book can be made to do work. Imagine tying a string to the book and attaching the other end to a nearby vase. Knock the book off the shelf, and watch it do work on the

KINDS OF ENERGY

Nine different kinds of energy are illustrated here, from the potential energy of the weights wound up in the clock to the awesome nuclear energy of an exploding atomic bomb, which releases vast quantities of heat, light, and sound.

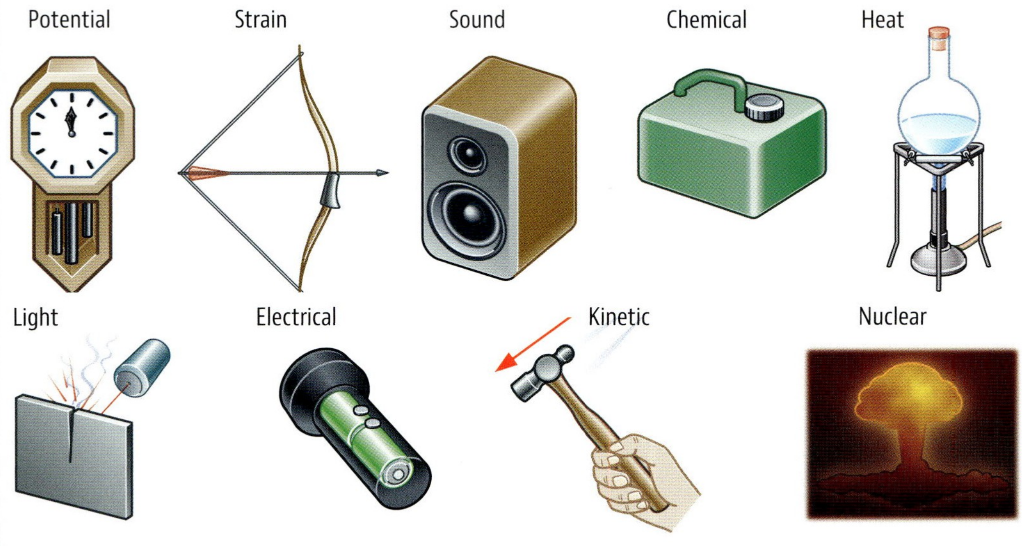

The water stored behind this dam represents a huge reserve of potential energy. As the water falls, the potential energy is converted to kinetic energy that can be made to do work by turning turbine blades.

vase! The water stored in a reservoir behind a dam has potential energy that can do work turning turbines to make electricity.

Strain energy is similar in some ways. When you wind up a clock or pull a bow, you strain the material of the spring or the bow and store energy in it. The clock spring slowly unwinds to work the clock, and the bow

SCIENCE WORDS

Energy: The capacity to do work. There are various kinds, including kinetic energy, potential energy, and strain energy. Heat, light, and sound are also forms of energy. Energy is measured in joules.
Strain energy: The energy an object possesses because its structure is strained (such as a stretched rubber band).
Work: The energy used when a force moves an object or changes its shape. It is measured in joules.

rapidly straightens to speed an arrow to its target. Kinetic energy is the energy of motion, so it is the form of energy possessed by anything that is moving. A swinging hammer has kinetic energy that can do the work of knocking a nail into a lump of wood. A speeding truck has lots of kinetic energy, which is why it can cause so much damage if it accidentally crashes into something.

Heat and light are also forms of energy. Anything hot possesses heat energy that can be made to do work in machines such as steam turbines and automobile engines. Green plants use light energy to combine carbon dioxide and water to form sugar and oxygen in the process known as photosynthesis, and light brings about other chemical reactions that are utilized in photography. The energy of the light in a laser beam is strong enough to cut through metal.

Electrical energy is one of the most familiar types of energy. It is produced by batteries and by generators in power plants. Electrical energy can be used to do all kinds of work from, for example, powering flashlights to driving railroad locomotives.

Sound is also a form of energy but it is seldom used directly for this purpose, although ultrasound is used in medical equipment to break up kidney stones, and in industry to cut through metals. Prolonged exposure to very loud sounds can damage human hearing, sometimes permanently.

The chemical energy locked up in fuels is released when the fuels are burned to

SCIENCE WORDS

Fission: The splitting of a heavy atomic nucleus into two roughly equal parts, with the subsequent release of energy.
Fusion: The merging of two light nuclei to form a heavier one, with the subsequent release of energy.

An archer stretches his bow and takes aim, about to convert the strain energy stored in the bow into the kinetic energy of a speeding arrow.

People on a roller coaster switch between having mostly potential energy to mostly kinetic energy. They retain some kinetic energy at the top of the ride because they keep moving, like the ball in the bowl (see right).

produce heat or light. Chemical energy is also released in batteries (but more slowly), where it is converted into electricity.

The final type of energy that we need to know about is nuclear energy. This is produced by changes that take place in the nuclei of atoms. Energy is released in fission reactions when large nuclei, such as uranium, split into smaller ones. Nuclear energy also comes from fusion reactions in which light nuclei, such as hydrogen nuclei, combine to form heavier ones. Fusion reactions occur at the heart of the Sun and other stars, as well as in hydrogen bombs.

ROLLING BALLS

A ball rolling back and forth in a basin switches between having all potential energy (at the highest positions) and all kinetic energy (at the lowest position).

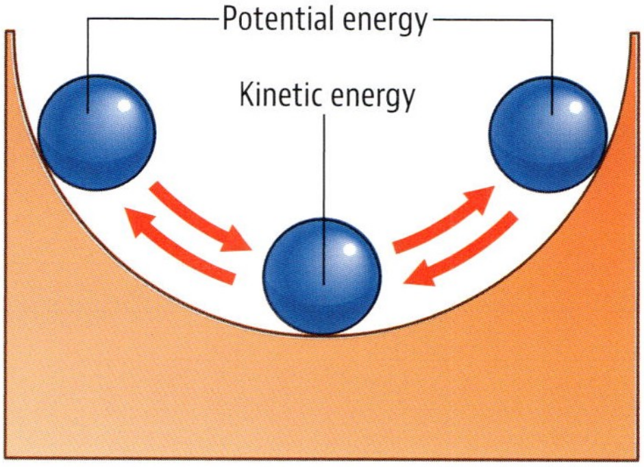

We noted earlier that energy cannot be created or destroyed, merely changed from one form into another. A few examples illustrate this point. The pendulum (see pages 34 to 35) is an energy converter. When the pendulum bob is at one end of its swing it has potential energy (because of its position). As it swings across it gains kinetic energy (because it is moving). The potential energy is converted to kinetic energy, which is changed back to potential energy at the other end of the swing.

A speeding bullet also has kinetic energy. When it hits a hard target, such as a wall, it stops. The kinetic energy is converted into heat energy, as well as some sound energy. When a meteorite enters the Earth's atmosphere, friction with the air heats it up (creating heat energy), and the meteorite ionizes atoms in the atmosphere, creating a brief streak of light (light energy), which is seen as a falling star.

JAMES JOULE

James Joule (1818–1889) was a British physicist who was best known for establishing connections between electricity and heat and between mechanical energy and heat. In 1840, he announced what is now called Joule's Law, which relates the electricity flowing in a wire and the wire's resistance to the amount of heat that is produced.

Three years later, Joule worked out how much heat is produced by a given amount of mechanical work, by measuring the heat produced when a container of water was stirred by paddles driven by falling weights. For several years Joule worked with William Thomson (Lord Kelvin, 1824–1907), and between them they discovered the Joule–Thomson effect, in which a gas is cooled when it escapes through narrow holes. This is the science behind the refrigeration process and how fridges work.

The sudden destructive release of chemical energy is evident in this detonation of explosives in a quarry.

The food we eat has chemical energy stored inside it. This energy drives our body processes, keeps us warm, and is used by our muscles whenever we do physical work or exercise.

Energy and Work

The energy content of food is generally measured in joules (or calories). And the joule is the unit used for measuring every other form of energy—for example, heat and mechanical energy. It is possible to compare energy sources in terms of joules. The joule is also the unit for measuring work. In fact, 1 joule is equal to a force of 1 newton moving through a distance of 1 meter. Whenever a force moves something, work is done. For example, if a book weighing 20 newtons (i.e., it has a mass of about 2 kg) is picked up off the floor and placed on a shelf

0.5 m high, the work done by the arm muscles is 20 * 0.5 = 10 joules. However, if a person carries three books up a flight of stairs that are 4 m high, the work done is 60 * 4 = 240 joules—making it much more tiring.

Work and Power

Power is the rate of doing work. If the three books were carried upstairs in 12 seconds, the power would be 240 ÷ 12 = 20 joules per second. But if the person ran up the stairs in just 4 seconds, the power would be 240 ÷ 4 = 60 joules per second.

Although these calculations give power in joules per second, in physics power has its own unit, the watt (1 watt = 1 joule per second), which was named after the Scottish engineer James Watt (1736–1819). The person running up the stairs converted energy at 60 watts—which is about the same power as a dim electric light bulb.

SCIENCE WORDS

joule (J): The derived SI unit of energy equal to the amount of work done when a force of 1 newton acts through a distance of 1 meter.

Kinetic energy: The energy an object possesses because it is moving.

Potential energy: The energy an object possesses because of its position (such as a weight that has been raised to a certain height above the ground).

SLOW AND FAST ELEVATORS

If these two elevators carry the same numbers of passengers, the express elevator needs four times as much power to move four times as fast as the slow elevator. It therefore needs a much larger electric motor to power its winding drum.

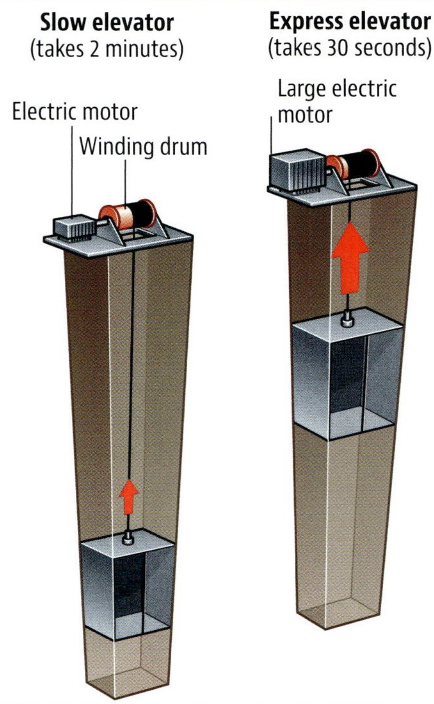

STABILITY AND EQUILIBRIUM

Normally, when a force acts on an object it makes it move in a straight line in the direction of the force. But in certain conditions a force may have no effect at all, or may cause the object to move in a circle, or even fall over.

The effect of a force on an object depends on how stable that object is. A cube-shaped box on a table top just sits there. It is perfectly stable and shows no tendency to move. If you tilt it slightly, raising one edge off the table, then let it go, it drops back down onto the table. In scientific terms, the box is said to be in stable equilibrium.

EQUILIBRIUM

The cube and the cone on its base are both in stable equilibrium, and each falls back onto its base if tilted. The ball and the cone on its side are both in neutral equilibrium and can roll. The cone on its point and the narrow cylinder are both in unstable equilibrium, and they will topple over if tilted far enough sideways.

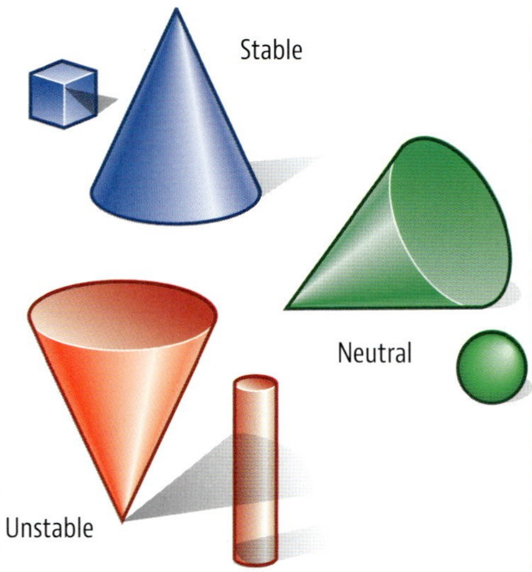

A cylinder lying on its side is slightly different. It will stay where it is put if it is placed on a level surface, but if given a slight push it will roll away. Such an object is said to be in neutral equilibrium. A narrow cylinder balanced on one end is different. Give it the slightest sideways tilt, and it topples over. It is in unstable equilibrium.

Center of Gravity

The point at which all of an object's mass appears to be concentrated is called its center of gravity (also sometimes called center of mass). In the cube, the center of gravity is right at the center. If you tilt the box, its center of gravity rises slightly. In the cylinder on its side, the center of gravity is halfway along its axis, which is the line joining the centers of each end. When the cylinder rolls, its center of gravity moves sideways, but does not move farther up or down.

The same cylinder on end has its center of gravity in the same place—halfway along its central axis. But this time, when the cylinder is tilted sideways, the center of gravity moves down slightly, and a vertical

SCIENCE WORDS

Center of gravity: Also called center of mass, the point at which an object's total mass appears to be concentrated and at which it acts.

Equilibrium: A state of physical balance. If an object in stable equilibrium is tilted, its center of gravity rises and when released, it falls back to its original position. If an object in neutral equilibrium is tilted, its center of gravity neither rises nor falls, and the object merely rolls. If an object in unstable equilibrium is tilted, its center of gravity falls, and the object will topple over.

This precariously balanced rock is the result of centuries of erosion. It is in unstable equilibrium. If it is tilted, it will topple over.

TRY THIS

Cheating the Force of Gravity

Whether or not an object can be balanced depends on its center of gravity. By setting up objects with unusual centers of gravity, the force of gravity can appear to be cheated.

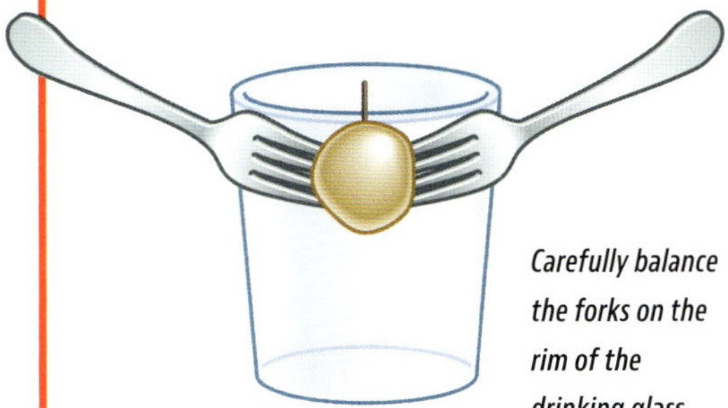

Carefully balance the forks on the rim of the drinking glass.

What to do

Make a small ball of modeling clay and carefully stick two forks in the ball at an angle. Press a toothpick into the side of the clay ball in the space between the two forks, as shown in the illustration above. Now balance the end of the toothpick on the edge of a drinking glass (you may have to move it slightly back and forth on the edge of the glass to find the best position). The result looks like magic!

This "trick" is possible because the center of gravity is in the space between the handles of the forks. As long as the toothpick passes through this point, the whole arrangement will balance with the center of gravity on the edge of the glass.

Another way of demonstrating this is to stick the forks (handles downward) into a lump of modeling clay molded onto a pencil near the pointed end. This arrangement will balance on the point of the pencil—try it on the base of the upturned glass. If you break off the pencil point and file a notch in the end with a nail file, you can balance it on a tightly stretched length of string, like a tightrope walker.

line through it meets the table at a position outside the cylinder's base. This makes the cylinder unstable, and it falls over.

In conclusion, the stability of an object depends on what happens to its center of gravity when it is tilted. If the center moves up, the object is in stable equilibrium. If the center moves sideways, it is in neutral equilibrium. But if the center moves down, the object is in unstable equilibrium.

Balancing Forces

A seesaw represents a different kind of equilibrium. Think of a seesaw pivoted at its center. It balances on the pivot (called a fulcrum in physics) because there are equal weights on either side. If unequal weights are added to each side—for example, a child at the end of one side and an adult at the end of the other—the seesaw tips downward on the heavier side. The unequal forces (the weights of the people are forces) produce a turning effect. This effect is called a turning moment, and its size is equal to the force multiplied by its distance from the fulcrum.

To balance the seesaw with the child and the adult, the turning moments have to be made the same. The only way to achieve this is for the adult to move nearer to the fulcrum so that the adult's weight multiplied by the distance to the fulcrum is the same as the product of the child's weight and the distance to the fulcrum. A turning moment is equal to a force (measured in newtons, N) multiplied by a distance (measured in meters, m), and its units are newton-meters, written as Nm. A monkey wrench is a good example of a practical use of turning moments. When changing a flat tire on a car, it is not possible to turn the wheel nuts using just your fingers—they are not strong enough. Using a wrench makes it easy because a

SCIENCE WORDS

Couple: The effect of two moments acting on an object in the same sense (i.e. both clockwise or both counter clockwise) at the same time.

Fulcrum: A pivot, as on a seesaw or where a lever pivots.

Moment: The turning effect (torque) produced when a force acts on an object, equal to the force multiplied by the perpendicular distance of its line of action to the pivot.

SIMPLE SEESAW

A seesaw is a simple example of turning moments. Children of equal weight balance at equal distances from the pivot. For an adult to have a ride, he or she has to sit closer to the pivot.

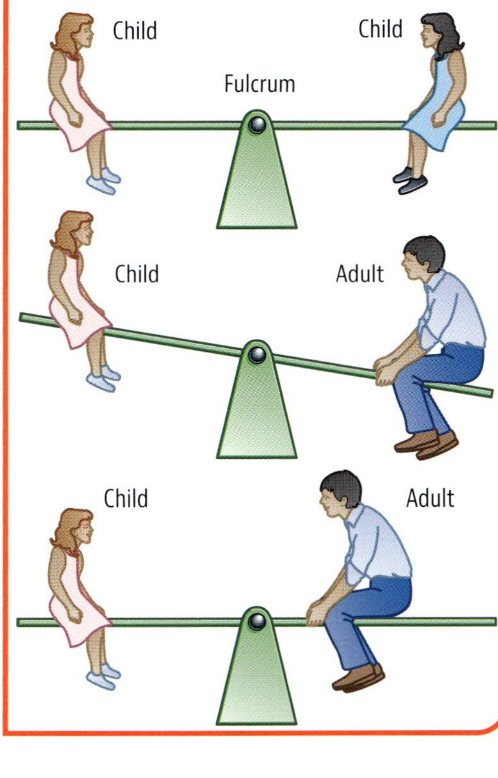

force of 100 N applied at right angles to the handle of a wrench 25 cm long produces a turning moment of 100 * 0.25 = 25 Nm. The crank between the pedal of a bicycle and the chain wheel acts in a similar way to produce rotation. Turning moments are also used to open doors. It takes more force to open a short wide door than a tall narrow one because the distance from the handle to the hinges is greater.

As long as this acrobat keeps his center of gravity (and that of the bowl on his head) over the cylinder below the board that is acting as a fulcrum, he will remain balanced.

Two Are Better than One

If there are two turning moments acting at the same point, the combined effect of both forces is called a couple. A familiar example is a faucet, which is turned by applying one moment on one side and an equal and opposite moment on the other side. Engineers sometimes use a long cylindrical wrench, called a socket wrench, for undoing stubborn nuts. It has a rod, termed a tommy bar, through one end of the cylinder. The other end goes over the nut, and the engineer applies a turning force to each end of the tommy bar. The turning effect of the couple unscrews the nut. A screwdriver is another example of a tool that uses couples. A screwdriver with a fat handle produces a greater couple and more turning force than one with a thin handle.

WRENCH AROUND

Applying a force F to the handle of a wrench at a distance **d** from the nut produces a turning moment **Fd**, also known as a torque.

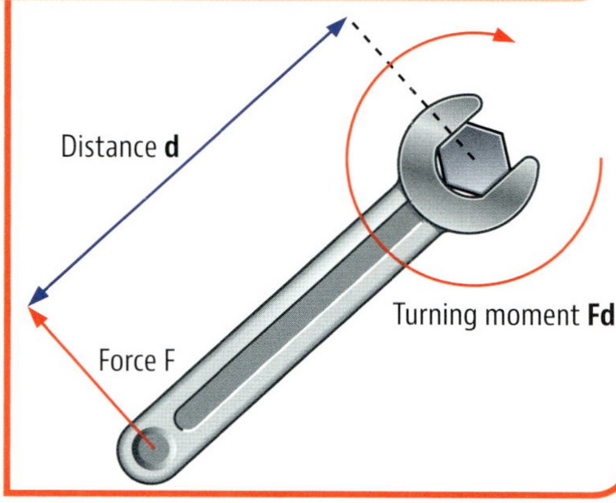

LOADS AND LEVERS

A lever is probably the simplest kind of machine, which can be defined as any device that provides a mechanical advantage. Even though they are simple, there are three very different kinds of lever, which have dozens of applications.

The ancient Greek scientist Archimedes is reputed to have once said, "Give me a lever long enough and somewhere to stand, and I will move the Earth." Archimedes (c. 287–212 B.C.E.) certainly made extensive use of levers in the ingenious machines he designed for his patron, the king of Syracuse.

SCIENCE WORDS

Fulcrum: A pivot, as on a seesaw or where a lever pivots.
Load: In a simple machine the output force (such as that applied by the effort).
Mechanical advantage: Also called force ratio, in a simple machine it is the load divided by the effort.

Classes of Lever

All kinds of lever have several things in common. They all involve a force, called the effort, that moves a load making use of a pivot, or fulcrum.

Archimedes' Earth-moving lever was a Class 1 lever, which works like a crowbar. Scissors or shears are a pair of Class 1 levers. The effort and load act in the same direction on opposite sides of the fulcrum.

In a Class 2 lever the effort and load are on the same side of the fulcrum and act in

THREE CLASSES OF LEVER

The three classes of lever are **Class 1**, with load and effort on opposite sides of the fulcrum; **Class 2**, with load and effort on the same side, but the load nearer the fulcrum; and **Class 3**, with load and effort again on the same side, but the effort nearer the fulcrum.

Class 1 lever

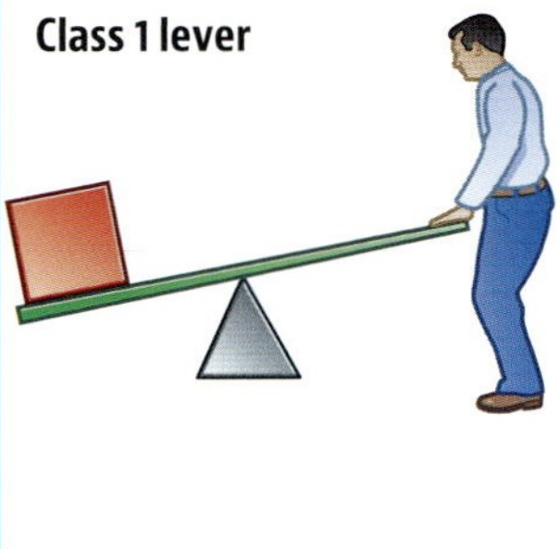

Class 2 lever

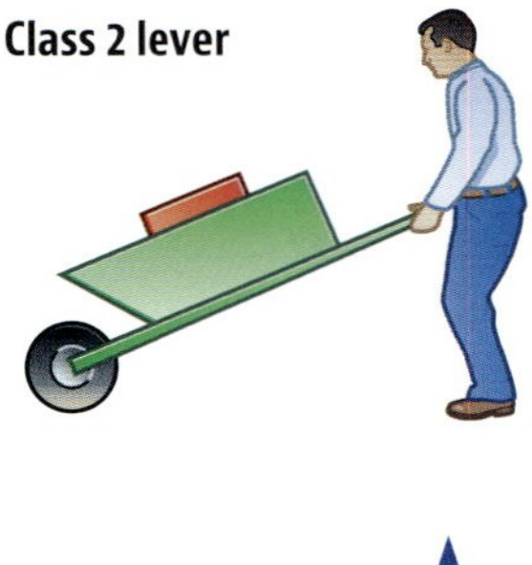

Class 3 lever

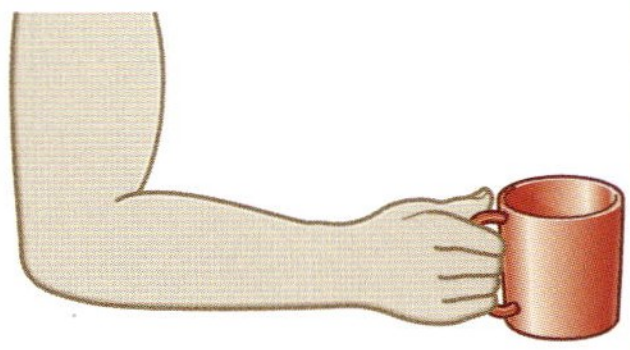

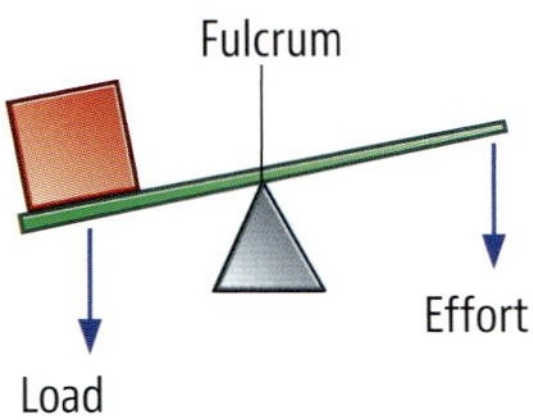

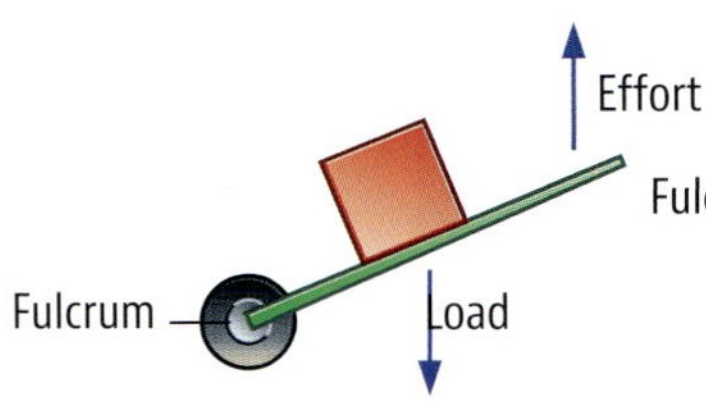

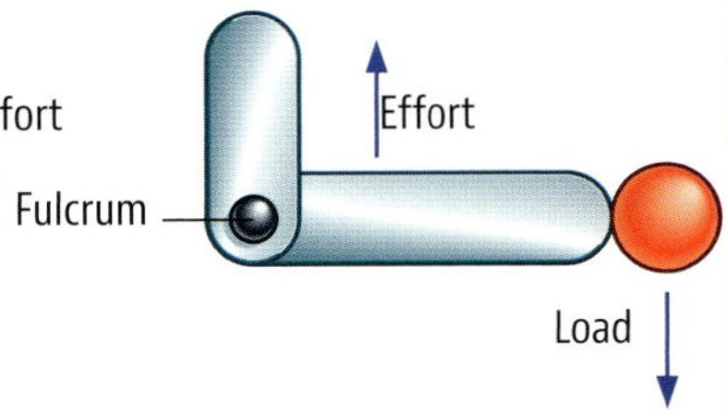

A pole-vaulter's pole acts as a lever to lift the athlete into the air. The vaulter is also helped by the springiness of the pole.

opposite directions. The load is nearer to the fulcrum than the effort. A wheelbarrow is a familiar example of a Class 2 lever.

Finally, a Class 3 lever also has the effort and the load on the same side and acting in opposite directions. This time, however, the effort is nearer the fulcrum than the load. The way your forearm works when you pick up something and the way tweezers and tongs work are examples of Class 3 levers in action. In fact, all movements of jointed bones in the human body involve levers of one type or another.

Mechanical Advantage

In a Class 1 lever, if the distance from the effort to the fulcrum is greater than the distance from the load to the fulcrum, then a small effort can move a large load. We say that the lever provides a mechanical advantage, which is defined as the load (output force) divided by the effort (input

A pair of garden shears, as with scissors, consists of a pair of Class 1 levers working together. The longer the handles, the greater the mechanical advantage.

force). This ratio is also sometimes called the force ratio, and for a Class 1 lever it is equal to the distance from the effort to the fulcrum divided by the distance from the load to the fulcrum. It also equals the distance the effort moves divided by the distance the load moves.

Most levers and other kinds of machine are useful because they have a mechanical advantage greater than 1: they increase forces. Imagine trying to pry the lid off a tin of paint using a coin as a kind of short crowbar. It acts as a Class 1 lever with a mechanical advantage of about 4. Because mechanical advantage is a ratio of two forces, it is a pure number and has no units. If this is not enough to remove the lid, the mechanical advantage can be greatly increased by using a screwdriver to pry off the lid—like a long crowbar. This will provide a mechanical advantage of up to 30, which should be more than enough to open it. The illustrations on these pages show other examples of various levers in action.

DUMP TRUCK

The hydraulic mechanism of this dump truck is an example of a Class 2 lever. Compare it with the wheelbarrow on page 46.

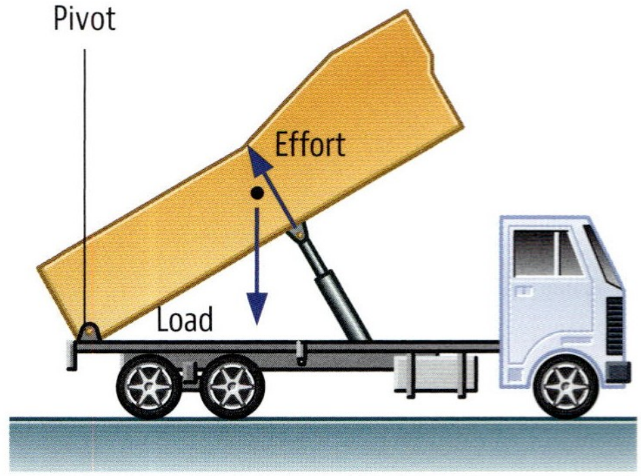

Wheel and Axle

A windlass is a device that uses a handle to turn a cylindrical drum to wind a bucket up from a well. A capstan used to pull up the anchor chain on a ship is another example. In scientific terms, such a device is called

A dump truck tipping its load is an everyday example of a Class 2 lever in action.

a wheel and axle. It is a variety of Class 1 lever in which the effort can be applied continuously. The input force is applied at the rim of the wheel, and the output force acts at the rim of the axle. If the wheel has a radius of R units and the axle has a radius of r units, the mechanical advantage is R divided by r. A car's steering wheel is another everyday example of a wheel and axle.

Mechanical Efficiency

All these devices that use levers are examples of simple machines. Some perform better than others—that is, some machines are more efficient than others. Efficiency is the energy (or power) produced by a machine, the useful work done, divided by the energy (or power) it consumes. Efficiency is usually expressed as a percentage and is always less than 100 percent since no machine is perfect.

In practice, there is usually a difference between any machine's theoretical mechanical advantage and the actual mechanical advantage. The ratio of these two—actual divided by theoretical—is also a measure of efficiency. A simple Class 1 lever is one of the most efficient machines, with an efficiency approaching 100 percent. Other simple machines, such as a screw (see page 52), are extremely inefficient.

LEVERING UPWARD

The cork lifter, or bottle opener, also employs a pair of Class 1 levers. The load is very close to the fulcrum, producing a large mechanical advantage that should shift even the most stubborn cork.

WHEEL AND AXLE

A wheel and axle are used in a windlass for raising a bucket from a well. This can provide a large mechanical advantage.

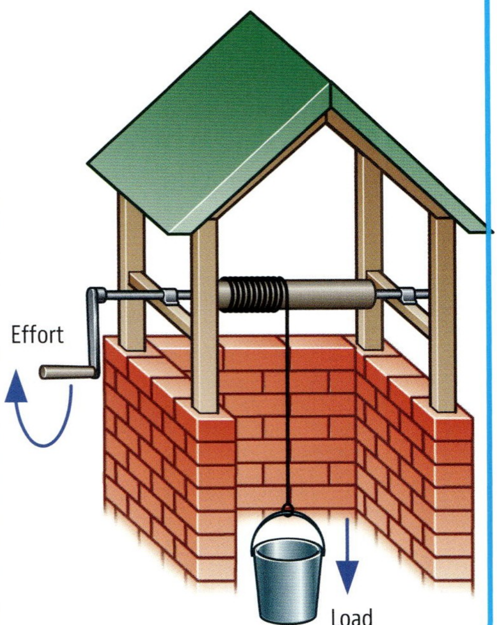

INCLINES AND FRICTION

It is easier to push a load up an incline than to lift it directly upward. The slope is called an inclined plane, and it is another simple machine that provides mechanical advantage. Without the slope, the pyramids of ancient Egypt would not have been built, and screws and bolts would not work.

Given the choice between a steep path straight up a hill or a gentle slope winding upward around it, most people would choose the gentler slope. Similarly, it is easier to climb a set of stairs than a vertical ladder. We do not think of such simple devices as being machines, but they are to a physicist. They are examples of an inclined plane.

Most machines provide a mechanical advantage that is always greater than 1. For an inclined plane, the mechanical advantage is the load (a downward force) divided by the effort (the force pushing the load up the slope), which is equal to the

The spiral staircase around the outside of this observation tower makes it easier to climb to the top than it would be if the tower just had a straight ladder.

SCIENCE WORDS

Friction: A force that prevents or slows the movement of one surface against another surface.
Inclined plane: A simple machine consisting of a ramp; the effort is used to push a load up the ramp. A wedge used to split things is also an inclined plane.
Mechanical advantage: Also called force ratio, in a simple machine it is the load divided by the effort.

length of the plane divided by the height of the slope. A wedge is a simple application of the inclined plane. Imagine driving a wedge under the edge of a heavy block. As the wedge moves in, it gradually lifts the block. This is exactly similar to pushing a block up an inclined plane, and the mechanical advantage is equal to the length of the wedge divided by its maximum thickness. Wedges have many uses, from splitting logs and rocks to forming the cutting part of an ax or chisel. The ancient Egyptians are thought to have built huge earth ramps—inclined planes—spiraling around the pyramids while they were building them. Slaves hauled large blocks of stone weighing many tons up the ramps, probably on rollers to reduce friction between the blocks and the ramp. When the pyramids were finished, the ramps were dismantled and the earth was removed to reveal the completed structures.

Spiral Inclined Planes

Winding a wedge (an inclined plane) around a cylinder can create much smaller spiral ramps. The result is a screw thread. When a screw is rotated in a block of wood, its threads cut into the wood and draw in the screw. Screws are tapered, but parallel-sided bolts work in the same way. The long narrow wedge that forms the thread makes a large mechanical advantage. Its value depends on the pitch of the screw, which is the distance it travels forward in one complete rotation. This, in turn, is equal to the distance between the screw's threads.

Overcoming Friction

A screw remains in a piece of wood because of friction. Friction is a force that prevents objects in contact from moving. Without it a screw would unscrew itself. But in the moving parts of machines friction is a nuisance, representing a waste of

INCLINED PLANE

It is easier to push a load up an inclined plane than to lift it up vertically. The mechanical advantage equals the length of the ramp (L) divided by its height (h).

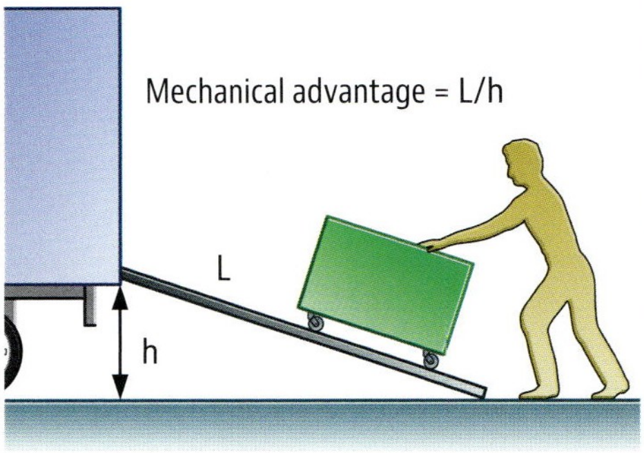

There is low friction between a skater's skates and the ice. When the foot is angled outward and the knee extended, the friction between the inside edge of the blade and the ice enables the blade to grip into the ice. High friction is used to stop on ice skates. More downward pressure creates drag, which removes the kinetic forward energy and converts it into heat and sound.

SCREWS AND JACKS

The mechanical advantage of a screw depends on its pitch, which is the distance between its threads. Rotating a screw enables it to lift things. This is the principle of the jackscrew, as used for lifting an automobile to change a tire.

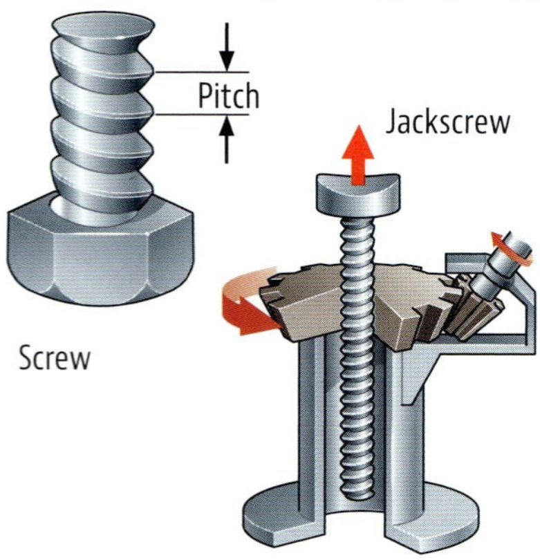

STATIC FRICTION

Static friction—the force needed to start an object sliding along a surface—depends on the object's weight, not on its shape or the area of contact between the object and the surface. All the objects shown here have the same weight (W), and static friction is the same for all of them. Lubricants reduce friction by introducing a layer of softer substance between the object and the surface.

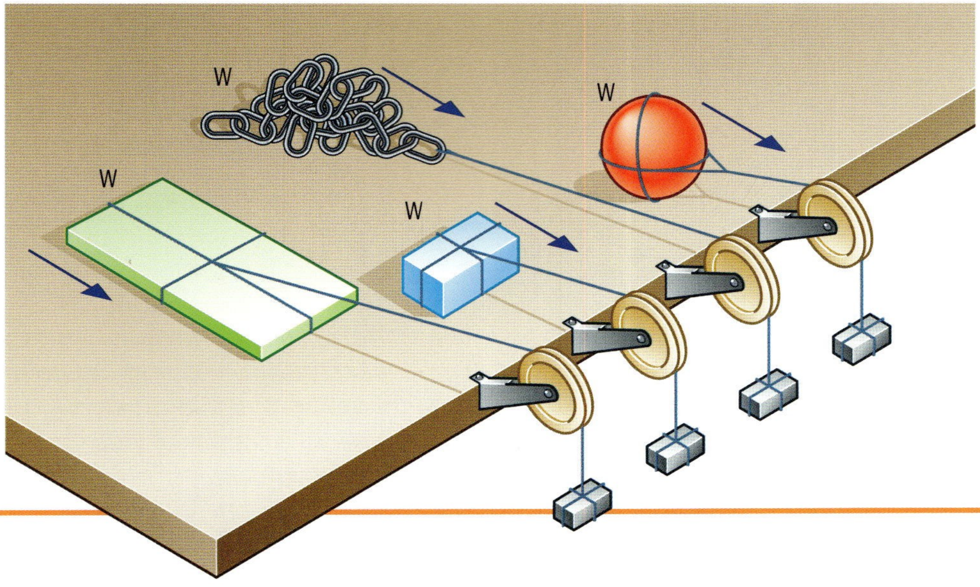

A technician using an angle grinder with a friction blade to cut through sheet steel. This type of saw uses its high speed of rotation to generate heat through friction in order to soften the metal in front of the blade, making it much easier to cut.

energy—for example, it soaks up about half the power of an automobile engine. Oil, grease, and other lubricants between the moving parts are designed to reduce friction.

Superfluidity

A strange low-temperature effect is called superfluidity. This term describes a fluid, such as liquid helium a few degrees above absolute zero (–460°F, –273°C), that flows without friction. Placed in a container, liquid helium flows vertically up the sides, over the lip, and escapes. On a slope, it flows uphill against the force of gravity. However, to date scientists have found very few practical uses for such superfluids.

TRY THIS

Keep on Rolling

Some ancient peoples moved huge heavy objects without any modern equipment. More than 3,000 years ago, the ancient Egyptians built the pyramids from large blocks of stone that were probably dragged by hundreds of slaves. To make the blocks move more easily, it is thought they used rollers.

What to do

Put a book on a table. Make a loose loop of string and put it around the book as shown in the illustration below. Tie a rubber band to the string. Now pull the book along by the rubber band toward the edge of the table. The amount the rubber band stretches is a measure of how much force you have to use. Now arrange six round pencils or marker pens in rows under the book so that they will roll toward you. Pull the book again using the rubber band. You will notice how you do not have to use so much force this time.

The force you have to overcome to get the book moving is friction. Sliding friction, as between the book and the table, is greater than the rolling friction between the pencils (or pens) and the table. Try placing a second book on top of the first one. How hard is it to move the two books?

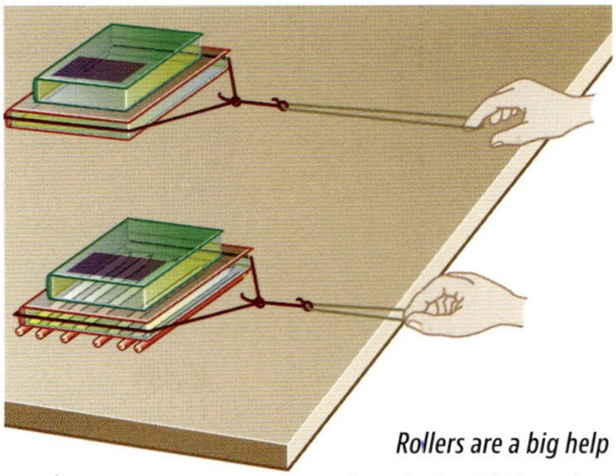

Rollers are a big help in reducing friction when moving a heavy load.

PULLEYS AND GEARS

Lifting heavy loads was a problem for ancient peoples, whose only help was the inclined plane and, later, the jackscrew. However, the problem was solved with the invention of pulleys. Later, gears were used to control the output of rotating machinery.

Most people today think of a machine as a useful device with wheels, gears, and other rotating parts. In many ways this is correct, although some modern machines (such as an electrical transformer) have no moving parts. We have seen levers and other simple machines on the previous pages, and nearly all of them were devised to help lift a load. One of the best lifting machines is a pulley.

Pulleys and Pulling

The simplest pulley has a rope passing over a single grooved wheel, like the type a farmer might use to haul a bale of hay up to a hayloft. In fact, a single pulley is not really a machine at all. Its mechanical advantage (the distance moved by the effort divided by the distance moved by the load) is 1, so there is no real advantage at all. And it cannot be used for lifting anything heavier than the person pulling on the rope. The single pulley simply changes the direction of a force—a downward pull on the rope lifts up the load.

However, two or more pulleys together give a mechanical advantage of 2, three pulleys together give a mechanical advantage of 3, and so on. The number of pulley wheels, or more precisely the number of ropes between them, produces the mechanical advantage of multiple pulleys, usually known as a block and tackle. With three wheels, however, the effort has to move three times as far as the load is lifted. As a result, a lot of rope has to be pulled through the pulley block to raise the load a small distance. Such pulleys came into their own in the days of sailing ships, when they were used to haul up the heavy weight of large canvas sails. They still have many uses today, especially in the large cranes that are used for lifting loads on construction sites and at shipyards.

SCIENCE WORDS

Block and tackle: A multiple pulley, one with two or more pulley wheels.
Machine: A device that allows one force (the effort) to overcome another (the load).
Pulley: A simple machine consisting of a fixed grooved wheel with a rope running around it. A mechanical advantage of more than 1 can be achieved only by using two or more pulleys together.

Belts and Gears

Early sources of power included water wheels and windmills. People needed a way of transferring the rotation of a shaft to other machines, such as millstones for grinding grain. One method was to link a wheel on the driven shaft to a wheel on the other machine using a rope or belt. Ropes ran on grooved wheels, and belts ran over wheels with flat rims. Belt drives were still in use long after the invention of the steam engine for connecting this new source of power to mechanized looms and metalworking machinery. The fan belt on an automobile, which enables the engine to turn the shaft of the electric generator, is a modern example of a belt drive.

On this heavy-duty winding mechanism (right) the smaller gear in front turns faster than the larger gear it drives.

By varying the sizes of the different wheels—the drive wheel and the driven wheel—the speed of rotation could be changed. A large drive wheel turns a smaller driven wheel faster, whereas a small drive wheel turns a larger driven wheel slower. Both wheels rotate in the same direction, unless the belt has a half turn like a figure eight. Larger and smaller wheels were also used when gears began to replace belt drives. Gears, also called cogwheels or cogs, are

SINGLE AND MULTIPLE PULLEYS

A single pulley only changes the direction of the pull on the rope. A multiple pulley, or block and tackle, provides a mechanical advantage. With four pulleys and four ropes between them the mechanical advantage is 4. The same pull **P**, as with the single pulley, will lift a load **W** that is four times as heavy.

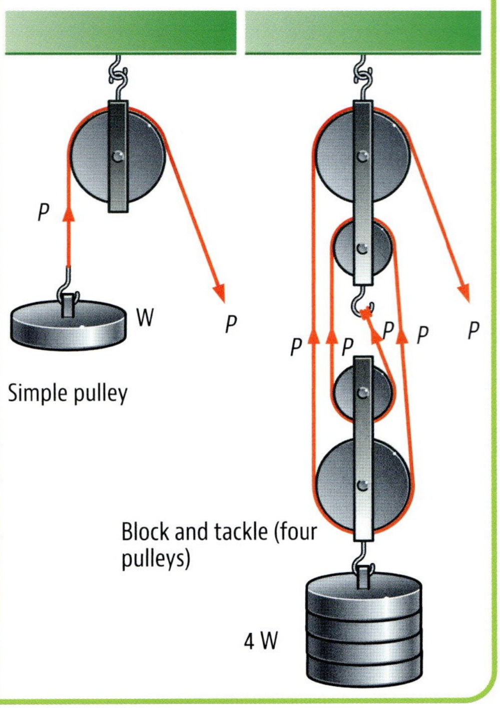

Mobile cranes dominate the skyline at a Shipyard, where they are used to lift parts and ships in and out of dry docks. All the lifting movements are achieved using pulleys.

toothed wheels, usually on parallel shafts, positioned so that the teeth of one engage with the teeth of the other. Again, a large drive gear turns a smaller gear (called a pinion) faster, and a small drive gear turns a larger gear slower. The two gears rotate in opposite directions. To make the pinion gear rotate in the same direction as the drive gear, a free-running idler gear is introduced between the two.

Other Kinds of Gear

A pinion gear driving a straight rod with teeth cut in it, called a rack, produces a side-to-side movement. It is called a rack and pinion, and is used in the steering mechanism of an automobile. A gear cut like a screw thread, called a helical gear or worm gear, can turn another gear on a shaft at right angles to it. Another way of changing the direction of rotation through a right angle is to use bevel gears, which are cut on an angle. Bevel

gears are used, for example, in differential drives of automobiles to transmit power to two axles spinning at different speeds, such as in cornering. See (**d**), (**e**), and (**f**) in the illustration below for these three types.

DIFFERENTIAL GEAR/DRIVE

When a truck or car goes around a curve, the wheel on the outer side of the curve goes farther, and rotates more, than the wheel on the inside of the curve. This arrangement allows it to happen.

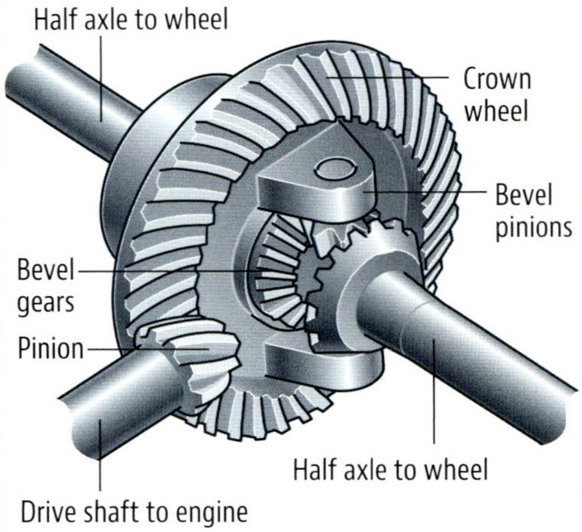

DIFFERENT KINDS OF GEAR

Shown here are (**a**) a drive gear turning a faster pinion in the opposite direction, (**b**) the use of an idler gear to keep the direction of rotation the same, (**c**) a small drive gear turning the driven gear slower, (**d**) a rack and pinion, (**e**) a helical gear, and (**f**) bevel gears.

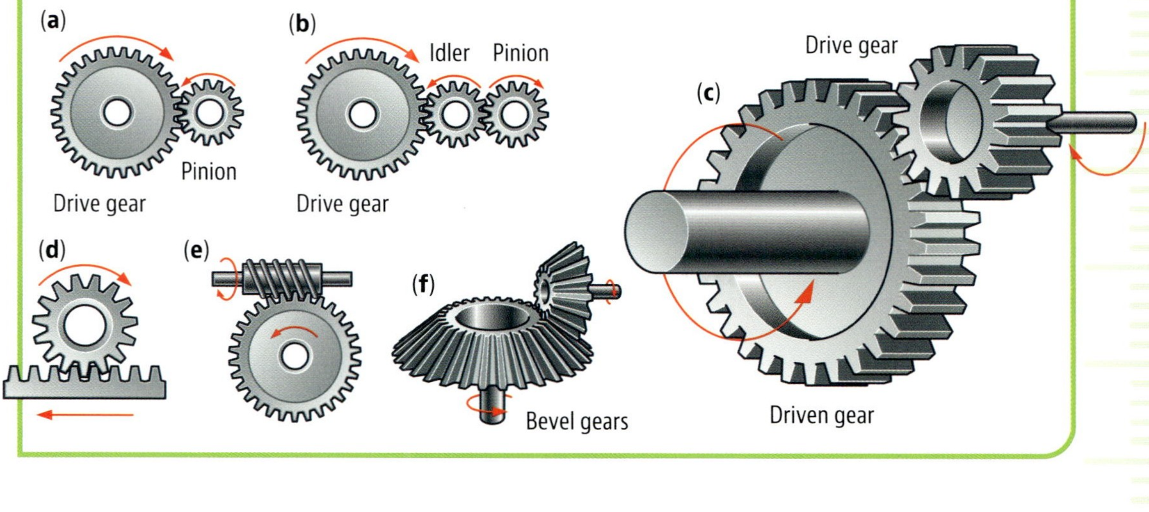

STRAINS ON SOLIDS

Solids can resist deforming forces much better than gases and liquids can. However, there is a limit even to the strength of solids—if you strain them enough, they will break. For lower strains they will return to their normal shape when the strain is removed.

Two key properties of most solids are strength and hardness. Nowhere are these properties used to better effect than in the construction industry. Over the ages, the chief permanent materials for construction have been stone, brick, concrete, and steel. They can be compared by considering how builders and engineers have solved the problem of building over a space to make a bridge.

The earliest material for bridges, not counting tree trunks and vines, was stone. The simplest stone bridge consists of a single slab across two supports, called a beam construction. But the length of such a span is limited to the longest piece of stone available. Masonry—square-cut pieces of stone—and its artificial imitator, brick, were next to be used. The secret of using these materials to stretch over long spans was the arch.

SCIENCE WORDS

Ductile: Describing a metal that can easily be drawn out to form wire.
Elasticity: The property of a solid that enables it to return to its original shape after it is stretched (that is, after it has been subjected to stress).
Strain: The change in shape of a solid object when it is subjected to a stress.
Stress: A force that tends to change the shape of a solid object, producing strain.

Many of the arches built in ancient Egypt, Greece, and Rome are still standing today. An arch is a sturdy structure because the materials that are used to make it are strong in compression—that is, when they are being squeezed. Each stone in an arch is in compression, and together they carry the load to the arch supports.

Concrete also has good compressive strength, and reinforced concrete, which has steel rods within its structure, is also strong in tension (when it is being stretched). A cantilever bridge is built into supports at each end. When it is loaded, a cantilever beam is in tension on the top and under compression beneath. Reinforced concrete is a good material for making bridges in this way. Steel is also strong in both tension and compression, and can be used for all types of

This three-tiered arch bridge, the Pont du Gard at Nîmes, is a Roman aqueduct. It was built from stone across the Gardon River in a valley in southern France.

BRIDGE CONSTRUCTION

Different kinds of bridges have different compressive forces. The decks of the beam **(a)** and cantilever **(b)** are also under tension. In an arch **(c)** each stone is compressed, and the load is carried to the arch supports.

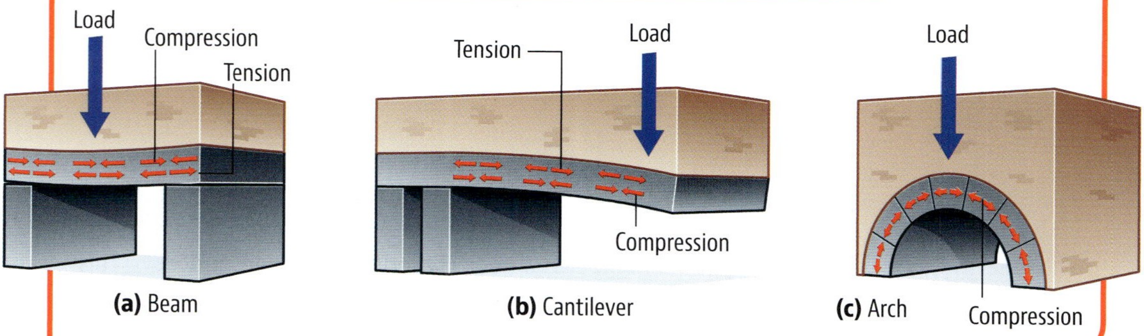

construction. There is a practical limit to the length of a steel girder, but this limitation is overcome by making trusses. Several short girders are joined together to make a rigid structure formed from triangles. Trusses are also much lighter in weight than any solid construction.

HOOKE'S LAW

As a metal sample is stretched, the load (stress) is proportional to the elongation (strain) up to the elastic limit. This is Hooke's Law. Beyond the yield point the sample stretches rapidly until it snaps at the breaking point.

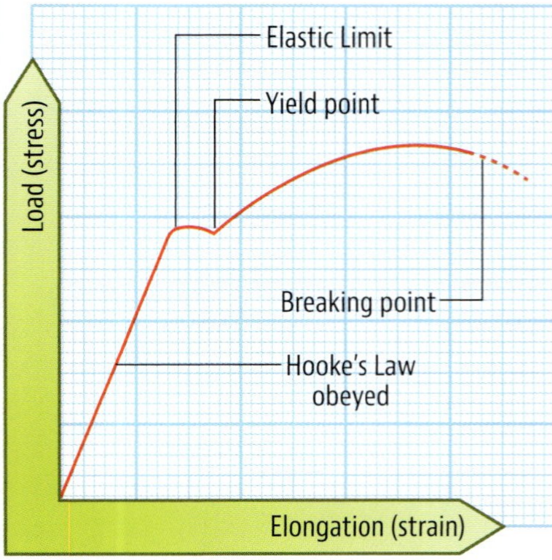

Elastic Metal

One property of steel and most other metals is elasticity. In scientific terms, an elastic material is one that stretches under tension, but returns to its original length when the tension is released. This property is very obvious with rubber and some plastics, which are collectively known as elastomers. But metals are also elastic, up to a point. That point is called the yield point. Imagine adding weights to the end of a length of wire fixed to a support at its upper end. When more weights are added, the wire stretches. If the weights are removed, the wire returns to its original length. In fact, the stress on the wire (the weight) is proportional to the strain (the amount it stretches). This relationship is known as Hooke's Law, after the English scientist Robert Hooke, who formulated the theory in 1676.

If more weights are added to the end of the wire, a point will be reached—called the elastic limit—where the wire does not return to its original length. It is permanently stretched. Beyond the elastic limit, at the yield point, the wire continues to stretch with very little extra load until it finally breaks.

ROBERT HOOKE

The English astronomer, physicist and engineer Robert Hooke (1635–1703) was the son of a clergyman. One of his most important insights in astronomy was his theory that the planets are held in their orbits by gravity. He proposed this theory in 1664, and it became the subject of a dispute with Isaac Newton (1642–1727), who also claimed to have had the idea. Hooke's major contributions to physics were the law that bears his name (see above right) and his discovery that most materials expand on heating. He also built a compound microscope for studying microscopic animals and plant cells, publishing his findings in the work *Micrographia* (1665). The many other Hooke inventions included the "universal joint", a device coupling a rotating shaft to another shaft at an angle.

Wire for Strength

A metal that is ductile can be pulled out to form wire. That is usually done in stages by pulling the wire through a series of dies. Each die is slightly smaller than the previous one, which gradually makes the wire thinner.

There are many practical uses for wire, particularly as a conductor of electricity. Wire is also used in construction. Several strands of wire can be twisted together to make a cable, which is stronger than a single strand of the same thickness. Cables are used in suspension bridges, in which two or more long cables hang in an arc between two towers. Thinner cables hang down from them to support a deck, which may be a roadway or a railroad.

Alloys for Hardness

As well as their density, most metals are also known for their hardness. This property results from the tight packing of the atoms in the crystals that form metals. Soft metals, such as aluminum and iron, can be made hard by alloying them with other elements, which modifies their crystal structure. Aluminum is mixed with magnesium or copper, and iron is alloyed with carbon to produce steel.

KINDS OF STRESS

The three main kinds of stress are compression, when a material is squeezed; tension, when it is stretched; and shear, when the top and bottom are pulled in opposite directions.

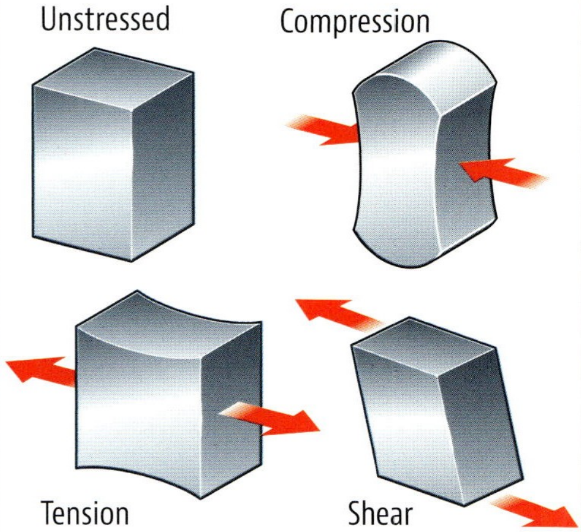

WINDING WIRE

Wire is made by pulling a thin metal rod through a series of dies, each one slightly smaller than the previous one.

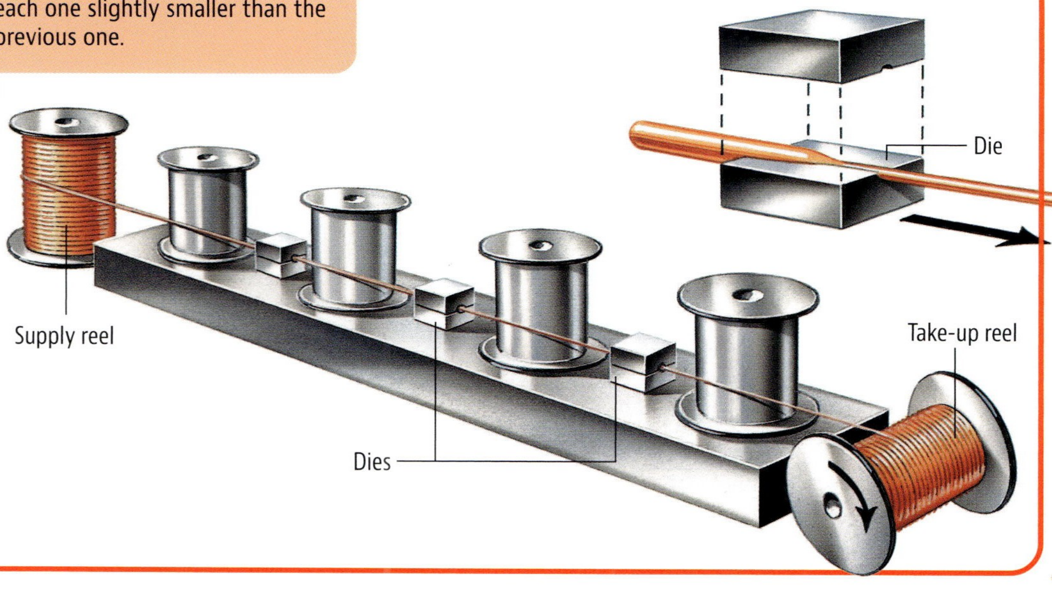

GLOSSARY

Acceleration The rate of change in a moving object's velocity.

Air resistance A force, also called drag or wind resistance, which resists the movement of an object passing through the air. It is overcome by streamlining.

Atomic number The number of protons in an element's nucleus. The atomic number equals the number of electrons in the normal atom and is the element's numerical position in the periodic table.

Calutron A large mass spectrometer used for separating the isotopes of uranium.

Center of gravity Also called center of mass, the point at which an object's total mass appears to be concentrated and at which it acts.

Centrifugal force A force that causes an object moving in a circular path to move out and away from the center of its path, in opposition to centripetal force.

Centrifuge A machine with a fast-spinning chamber that creates strong "artificial gravity" to separate materials of different densities.

Centripetal force The force that acts inward to keep an object moving in a circle.

Couple The effect of two moments acting on an object in the same way (i.e. both clockwise or both counter clockwise) at the same time.

Ductile Describing a metal that can easily be drawn out to form wire.

Elasticity The property of a solid that enables it to return to its original shape after it has been stretched (that is, after it has been subjected to stress).

Electron A negatively charged elementary particle found in every atom.

Energy The capacity to do work. There are various kinds, including kinetic energy, potential energy, and strain energy. Heat, light, and sound are also forms of energy.

Equilibrium A state of physical balance.

Fission The splitting of a heavy atomic nucleus into two roughly equal parts, with the release of much energy.

Force An influence that changes the shape, position, or movement of an object.

Friction A force that prevents or slows the movement of one surface against another.

Fulcrum A pivot, as on a seesaw or where a lever pivots.

Fusion The merging of two light nuclei to form a heavier one, with the release of energy.

Gas diffusion A technology used in the enrichment of materials used in nuclear reactors and nuclear weapons.

Inclined plane A simple machine consisting of a ramp. A wedge used to split things is also an inclined plane.

Inertia The property of an object that makes it tend to resist being moved or, if moving, to resist a change in direction. It is a consequence of the first of Newton's laws of motion.

Ion An atom or molecule that has lost or gained one or more electrons, so gaining an electric charge.

Isotope Any of the varieties of a chemical element that are chemically identical to one another, but whose atoms differ in their relative atomic mass.

joule (J) The derived SI unit of energy equal to the amount of work done when a force of 1 newton acts through a distance of 1 meter.

Kinetic energy The energy an object possesses because it is moving.

Load In a simple machine, the output force (such as that applied by the effort).

Luminous intensity The light-emitting power of a source of light.

Machine A device that allows one force (the effort) to overcome another (the load).

Mass The amount of matter in an object.

Mass spectrograph An instrument that separates fast-moving ions in a beam according to their mass.

Mechanical advantage Also called force ratio, in a simple machine it is the load divided by the effort.

Moment The turning effect (torque) produced when a force acts on an object, equal to the force multiplied by the perpendicular distance of its line of action to the pivot.

Momentum The mass of an object multiplied by its velocity in a straight line.

Newton (N) The derived SI unit of force. The force required to give a mass of 1 kilogram an acceleration of 1 meter per second per second.

Potential energy The energy an object has because of its position (such as a weight raised to a certain height above the ground).

Pulley A simple machine consisting of a fixed grooved wheel with a rope running around it.

Relative atomic mass The mass of an atom expressed in atomic mass units (amu). The atomic mass unit is defined as one-twelfth of the mass of the common isotope of carbon and is approximately equal to the mass of the hydrogen atom.

Scalar A quantity that has magnitude, but (unlike a vector) no specified direction. Examples of scalars are speed and mass.

SI units System of units used internationally in science. There are seven base units (ampere, candela, kelvin, kilogram, meter, mole, and second) and various derived units, which are combinations of base units.

Strain The change in shape of a solid object when it is subjected to a stress.

Strain energy The energy an object possesses because its structure is strained (such as a stretched rubber band).

Stress A force that tends to change the shape of a solid object, producing strain.

Terminal speed The maximum speed at which an object falls under the influence of gravity.

Thermocouple Temperature-measuring device comprising two wires of different metals joined at their ends.

Vector A quantity that has magnitude and (unlike a scalar) direction. Examples of vectors are acceleration and velocity.

Weight The force with which a mass is drawn toward the Earth (by the force of gravity).

Work The energy used when a force moves an object or changes its shape, measured in joules.

FURTHER RESEARCH

Books

Clark, John O E. *The Basics of Mechanics.* New York, NY: Rosen Young Adult, 2024.

Green, Dan. *The Science Book* (DK Big Ideas). New York, NY: DK Publishing, 2025.

Macaulay, David. *The Way Things Work: Newly Revised Edition.* New York, NY: Clarion Books, updated edition 2023.

Web Sites

How Stuff Works–Physical Science
howstuffworks.com/physical-science
Topics on all aspects of physics.

Nasa Glenn Research Center
https://www1.grc.nasa.gov/beginners-guide-to-aeronautics/newtons-laws-of-motion/#what-are-newtons-laws-of-motion
Newton's Laws of Motion explained, including a short movie.

PhysLink.com
PhysLink.com
www.physlink.com/
Physics education, research, and reference.

INDEX